Lecture Notes in Economics and Mathematical Systems

540

Springer
Berlin
Heidelberg
New York
Hong Kong
London
Milan
Paris
Tokyo

Holger Kraft

Optimal Portfolios with Stochastic Interest Rates and Defaultable Assets

Author

Dr. Holger Kraft
Fraunhofer ITWM
Gottlieb-Daimler-Straße
67663 Kaiserslautern
Germany

Cataloging-in-Publication Data applied for

Bibliographic information published by Die Deutsche Bibliothek.
Die Deutsche Bibliothek lists this publication in the Deutsche Nationalbibliographie; detailed bibliographic data is available in the Internet at <http://dnb.ddb.de>.

ISSN 0075-8442
ISBN 3-540-21230-2 Springer-Verlag Berlin Heidelberg New York

Springer-Verlag is a part of Springer Science+Business Media

springeronline.com

Printed in Germany

Typesetting: Camera ready by author
Cover design: *Erich Kirchner*, Heidelberg

Printed on acid-free paper 55/3142/du 5 4 3 2 1 0

To my parents

Preface

This thesis summarizes most of my recent research in the field of portfolio optimization. The main topics which I have addressed are portfolio problems with stochastic interest rates and portfolio problems with defaultable assets.

The starting point for my research was the paper "A stochastic control approach to portfolio problems with stochastic interest rates" (jointly with Ralf Korn), in which we solved portfolio problems given a Vasicek term structure of the short rate. Having considered the Vasicek model, it was obvious that I should analyze portfolio problems where the interest rate dynamics are governed by other common short rate models. The relevant results are presented in Chapter 2.

The second main issue concerns portfolio problems with defaultable assets modeled in a firm value framework. Since the assets of a firm then correspond to contingent claims on firm value, I searched for a way to easily deal with such claims in portfolio problems. For this reason, I developed the elasticity approach to portfolio optimization which is presented in Chapter 3. However, this way of tackling portfolio problems is not restricted to portfolio problems with defaultable assets only, but it provides a general framework allowing for a compact formulation of portfolio problems even if interest rates are stochastic. To this end, Chapter 3 is also closely connected to Chapter 2. As mentioned above, defaultable assets in firm value models correspond to contingent claims on firm value. More precisely, they are mostly just barrier derivatives with respect to a curved boundary. Therefore, in Chapter 4 I prove formulae which are tailored to determine the value of debt and equity in those firm value models which I have considered. Finally, in Chapter 5, which is partly the result of work jointly undertaken with Ralf Korn, portfolio problems with defaultable assets are solved. Hence, apart from Chapter 1 which provides a survey of relevant results from the theory of stochastic differential equations and stochastic control, the other chapters can be assigned to at least one of my main issues "Optimal Portfolio with Stochastic Interest Rates" and "Optimal Portfolio with Defaultable Assets". The basis for these chapters are papers which I have written during recent years. For this reason, I have retained the presentation of the chapters, each of which starts with an introduction and ends with a conclusion, as self-contained entities.

Clearly, the preparation of this thesis would not have been possible without the support of many people. Above all I wish to thank Ralf Korn for his cheerful supervision and many fruitful discussions as well as his comments and suggestions on various parts of the text. For me it was a stroke of good luck that I attended his last lectures at the University of Mainz although, at that time, I did not know anything about stochastic control theory. Nevertheless, these lectures prepared the ground for my master thesis which I wrote under his supervision and this doctoral thesis. Further, I wish to thank Christian Hipp for taking over the duty of being the second referee. I also wish to thank my current and former colleagues Beate Jüttner-Nauroth, Martin Krekel, Anne-Kerstin Kampovsky, Gerald Kroisandt, Tin-Kwai Man, Olaf Menkens, Marlene Müller, and Frank Seidel. Besides, I wish to thank Mogens Steffensen for fruitful discussions which helped me for example to solve the portfolio problem considered in Subsection 5.2.3. Finally, I thank my former mentors Siegfried Trautmann and Hans-Jürgen Schuh for sharpening my understanding of finance and stochastics.

Special thanks are due to my partner, Maria Seemann, for her patience and support.

Wiesbaden, December 2003 *Holger Kraft*

Contents

1

Preliminaries from Stochastics

Portfolio problems in continuous time can be interpreted as control problems. To this end, in this chapter we sum up results of the theory of stochastic control which are relevant to our further considerations.[1]

We look at the dynamics of an n-dimensional Ito process Y which is said to be the state process. This process depends on an m-dimensional Brownian motion W. Starting with initial value $Y(t_0) = y_0$ this process can be controlled by a process $u(\cdot)$ which, in general, is involved in the drift and the diffusion process of the controlled stochastic differential equation (CSDE)[2]

$$dY(t) = \Lambda(t, Y(t), u(t))dt + \Sigma(t, Y(t), u(t))dW(t) \tag{1.1}$$

which models the dynamics of the state process. Considering a portfolio problem, Y describes the dynamics of the investor's wealth. The control $u(\cdot)$ corresponds to the allocation of the investor's wealth to stocks, bonds, and the money market account.

From the theory of stochastic differential equations it is known that the coefficients of a stochastic differential equation (SDE) have to be chosen carefully to ensure the existence and uniqueness of a solution. This problem becomes more difficult for SDE with stochastic coefficients such as (1.1). Consequently, one cannot allow for arbitrary coefficients and controls.

[1] Most results in this chapter are taken from Fleming/Soner (1993), pp. 125ff, Korn/Korn (2001), pp. 223ff, and Korn/Kraft (2001).

[2] Note that stochastic differential equations and their solutions are only determined almost surely. In the following this fact will not be mentioned explicitly.

The chapter is organized as follows. In Section 1.1 we recapitulate relevant results on SDEs. Keeping these results in mind in Section 1.2 we give the above CSDE a precise mathematical meaning and formulate a so-called verification theorem in its classical form.[3] This theorem is an important tool to solve control problems. However, it can be applied only under certain growth and Lipschitz conditions which are sometimes not satisfied in our portfolio problems. Nevertheless, if CSDE (1.1) has a linear form, one can prove a modified verification theorem where these conditions are replaced by more general requirements.

1.1 Stochastic Differential Equations

In general, it is not possible to compute a solution of an SDE explicitly. An important exception are linear SDEs.

Let us consider a complete probability space $(\Omega, \mathcal{F}, P)$. The m-dimensional process $\{(W(t), \mathcal{F}_t)\}_{t\in[0,\infty)}$ denotes a Brownian motion defined on this space and $\{\mathcal{F}_t\}_{t\in[0,\infty)}$ the respective Brownian filtration.[4] Unless otherwise stated, every adapted or a progressively measurable process is understood to be adapted or progressively measurable with respect to the Brownian filtration. Let $|\cdot|$ denote the Euclidean or the operator norm.

Proposition 1.1 (Variation of Constants) *Assume that $x \in I\!R$ and A, $\tilde{A}$, B_j and $\tilde{B}_j$, $j = 1, \ldots, m$, are real-valued stochastic processes which are progressively measurable with respect to $\{F_t\}_t$ and meet the following integrability conditions*

$$\int_0^t \left(|A(s)| + |\tilde{A}(s)|\right) ds < \infty \quad P\text{-a.s., for all } t \geq 0,$$

$$\int_0^t \left(B_j(s)^2 + \tilde{B}_j(s)^2\right) ds < \infty \quad P\text{-a.s., for all } t \geq 0.$$

Then the linear inhomogeneous SDE

[3] See Fleming/Rishel (1975), pp. 159f, Fleming/Soner (1993), p. 163, and Korn/Korn (2001), pp. 229f.

[4] The Brownian filtration meets the usual conditions, i.e. it is right-continuous and $\mathcal{F}_0$ contains all P-null sets of $\mathcal{F}$. See Korn/Korn (2001), p. 18.

$$dX(t) = \Big(A(t)X(t) + \tilde{A}(t)\Big)dt + \sum_{j=1}^{m}\Big(B_j(t)X(t) + \tilde{B}_j(t)\Big)dW_j(t)$$

with initial condition $X(0) = x$ has the Lebesgue $\bigotimes P$-unique solution $\{(X(t), \mathcal{F}_t)\}_{t\in[0,\infty)}$ given by

$$X(t) = Z(t)\left(x + \int_0^t \frac{1}{Z(u)}\left(\tilde{A}(u) - \sum_{j=1}^{m} B_j(u)\tilde{B}_j(u)\right) du + \sum_{j=1}^{m}\int_0^t \frac{\tilde{B}_j(u)}{Z(u)} dW_j(u)\right),$$

where

$$Z(t) = \exp\left(\int_0^t \left(A(u) - 0,5|B(u)|^2\right) du + \int_0^t B(u)dW(u)\right)$$

is the solution of the homogeneous linear SDE

$$dZ(t) = Z(t)\Big[A(t)dt + B(t)'dW(t)\Big]$$

with $Z(0) = 1$.

A proof can be found in Korn/Korn (2001, pp. 55f).[5]

In the following we also look at nonlinear SDEs of the form

$$dX(t) = A(t, X(t))dt + B(t, X(t))dW(t) \tag{1.2}$$

with initial condition $X(0) = x_0 \in I\!R^k$, $k \in I\!N$, where A and B are allowed to be stochastic processes.[6] To give (1.2) a precise mathematical meaning, define $Q_0 := [t_0, t_1) \times I\!R^k$ for $0 \le t_0 < t_1 < \infty$. Let the coefficients be functions $A : \bar{Q}_0 \times \Omega \to I\!R^k$ and $B : \bar{Q}_0 \times \Omega \to I\!R^{k,m}$, $m \in I\!N$. Then we obtain the following theorem which can be found in Fleming/Soner (1993, pp. 397f).

Theorem 1.1 (Existence and Uniqueness) *For fixed $t \in [t_0, t_1]$ and $x \in I\!R^k$ let $A(\cdot, x, \cdot)$ and $B(\cdot, x, \cdot)$ be progressively measurable with respect to $\{\mathcal{F}_s\}_{s\in[0,\infty)}$ on $[t, t_1]$. Besides, with a constant $K > 0$ for all $s \in [t_0, t_1]$, $x, y \in I\!R^k$, the global Lipschitz conditions*

[5] See also Karatzas/Shreve (1991), pp. 360f.

[6] For notational convenience we mostly omit the dependency of A and B on ω.

$$|B(s,x) - B(s,y)| \leq K|x-y|,$$
$$|A(s,x) - A(s,y)| \leq K^2|x-y|,$$

and the growth conditions

$$0,5|B(s,x)|^2 \leq g(s)^2 + K^2|x|^2,$$
$$|A(s,x)| \leq h(s) + K^2|x|$$

are met where g and h are processes that fulfill the integrability conditions

$$\mathrm{E}\left(\int_t^{t_1} g(s)^\rho + h(s)^\rho \, ds\right) < \infty$$

for all $\rho > 0$.

Then (1.2) has a pathwise unique solution, i.e. for two arbitrary solutions $X^{(1)}$, $X^{(2)}$ we obtain $P(X^{(1)}(t) = X^{(2)}(t) \, \forall \, t \in [t_0, t_1]) = 1$.

Krylov (1980, p. 85) proved an estimate for the moments of the solution of (1.2):

Theorem 1.2 (Estimate of the Moments) *If the assumptions of Theorem 1.1 still hold, then for every solution of (1.2) and every $\rho \geq 2$ there exists a constant $N = N(\rho, K) > 0$ so that*

$$\mathrm{E}\Big(\big(\sup_{s\in[t,t_1]} |X(s)|\big)^\rho\Big)$$
$$\leq N|x_0|^\rho + N(t_1-t)^{0,5\rho-1} e^{N(t_1-t)} \mathrm{E}\left(\int_t^{t_1} \left(|x_0|^\rho + g(s)^\rho + h(s)^\rho\right) ds\right)$$

for all $t \in [t_0, t_1]$.

By imposing weaker requirements we can get rid of the Lipschitz conditions as the following theorem shows.

Proposition 1.2 (Existence/Uniqueness without Lipschitz Cond.) *Let a and b be real-valued, continuous functions on $\mathbb{R}_0^+ \times \mathbb{R}$. Suppose that there exist positive increasing functions h_a, $h_b : \mathbb{R}_0^+ \to \mathbb{R}_0^+$ with*

$$|a(s,x) - a(s,y)| \leq h_a(|x-y|), \tag{1.3}$$
$$|b(s,x) - b(s,y)| \leq h_b(|x-y|), \tag{1.4}$$

and $\int_{(0,\varepsilon)} \frac{1}{h_a(u)} \, du = +\infty$ as well as $\int_{(0,\varepsilon)} \frac{1}{h_b^2(u)} \, du = +\infty$ for all $\varepsilon > 0$. Besides, for $K > 0$ the growth condition

$$a^2(t,x) + b^2(t,x) \le K(1+x^2)$$

is met. Then the SDE

$$dX(t) = a(t,X(t))dt + b(t,X(t))dW(t), \quad X(0) = x_0,$$

has a pathwise unique solution, i.e. for two arbitrary solutions $X^{(1)}$, $X^{(2)}$ we obtain $P(X^{(1)}(t) = X^{(2)}(t)\ \forall\, t \in [t_0, t_1]) = 1$.

Remarks.

a) The existence result goes back to Skorokhod (1965, pp. 59ff). To prove existence, the conditions (1.3) and (1.4) are not required.

b) The uniqueness result is proved by Yamada/Watanabe (1971) and reported by Karatzas/Shreve (1991, pp. 291ff).

c) Actually, under the assumptions of Proposition 1.2 one can show that strong uniqueness and strong existence hold. We refer the reader to Yamada/Watanabe (1971), Yamada (1981), or Karatzas/Shreve (1991, pp. 291ff).

d) Proposition 1.3 covers volatility coefficients of the form $b(t,x) = \sqrt{x}$ because $|\sqrt{x} - \sqrt{y}| \le \sqrt{|x-y|}$ for x, $y \ge 0$. This inequality can be proven as follows: Without loss of generality fix $x > y > 0$. Then there exist u, $v > 0$ such that $x = u^2$, $y = v^2$. Obviously, $\sqrt{u-v} \le \sqrt{u+v}$ and therefore $u - v \le \sqrt{u-v}\sqrt{u+v}$. Hence, $|u-v| \le \sqrt{|u^2 - v^2|}$.

Additionally, a so-called comparison result for SDE may sometimes be helpful:

Proposition 1.3 (Comparison Result for SDE) *Suppose that for continuous, real-valued functions a_1, a_2, and b on $\mathbb{R}_0^+ \times \mathbb{R}$ the following processes are given by*

$$X^{(j)}(t) = X^{(j)}(0) + \int_0^t a_j(s, X^{(j)}(s))\, ds + \int_0^t b(s, X^{(j)}(s))\, dW(s),$$

$j = 1, 2$. Assume further that

- *b meets condition (1.4),*
- *$X^{(1)}(0) \le X^{(2)}(0)$,*
- *$a_1(t,x) \le a_2(t,x)$, $(t,x) \in \mathbb{R}_0^+ \times \mathbb{R}$,*
- *with a constant $K > 0$ either a_1 or a_2 satisfies*

$$|a_j(t,x) - a_j(t,y)| \le K|x-y|.$$

Then $$P(X^{(1)} \leq X^{(2)}) = 1.$$

A proof can be found in the book by Karatzas/Shreve (1991, p. 293).

1.2 Stochastic Optimal Control

We are now in a position to give the CSDE (1.1) a precise mathematical meaning. Let $[t_0, t_1]$ with $0 \leq t_0 < t_1 < \infty$ be the relevant time interval. A control $u(\cdot)$ is a progressively measurable process with $u(t) \in U$ for all $t \in [t_0, t_1]$, where the set $U \subset I\!R^d$, $d \in I\!N$, is assumed to be closed. As before, let $Q_0 := [t_0, t_1) \times I\!R^n$, $n \in I\!N$. Besides, it is assumed that the coefficients

$$\begin{aligned} \Lambda &: \bar{Q}_0 \times U \to I\!R^n, \\ \Sigma &: \bar{Q}_0 \times U \to I\!R^{n,m}, \end{aligned}$$

$m \in I\!N$, are continuous and for all $v \in U$ the functions $\Lambda(\cdot, \cdot, v)$ and $\Sigma(\cdot, \cdot, v)$ belong to $C^1(\bar{Q}_0)$.

For a fixed control $u(\cdot)$ the solution of the corresponding CSDE is denoted by Y^u.[7] To emphasize that the expected value of a solution at time $t \in [t_0, t_1]$ depends on the initial condition, we use the notation $\mathrm{E}^{t_0, y_0}(Y(t))$.

Definition 1.1 (Admissible Control) *A control* $\{(u(t), \mathcal{F}_t)\}_{t \in [t_0, t_1]}$ *is said to be admissible if*

(i) for all initial conditions $y_0 \in I\!R^n$ *the corresponding CSDE (1.1) with* $Y(t_0) = y_0$ *has a pathwise unique solution* $\{Y^u(t)\}_{t \in [t_0, t_1]}$,

(ii) for all $k \in I\!N$ *the integrability condition*

$$\mathrm{E}\left(\int_{t_0}^{t_1} |u(s)|^k \, ds\right) < \infty$$

is met, and

(iii) the corresponding state process Y^u *meets*

$$\mathrm{E}^{t_0, y_0}\left(\sup_{t \in [t_0, t_1]} |Y^u(t)|^k\right) < \infty.$$

---

[7] Strictly speaking, one should denote the process by $Y^{t_0, y_0, u}$ because the solution depends on the initial condition. However, for simplicity of notation, it is denoted by Y^u or briefly by Y.

$\mathcal{A}(t_0, y_0)$ denotes the set of all admissible controls with respect to the initial condition $(t_0, y_0) \in Q$.

From the results of the previous section it is evident that one can put specific Lipschitz and growth conditions on the coefficients of the CSDE to make sure that the CSDE has a unique solution for an admissible control. For instance, it can be assumed that

$$|\Lambda_t| + |\Lambda_y| \leq C, \qquad (1.5)$$
$$|\Sigma_t| + |\Sigma_y| \leq C,$$
$$|\Lambda(s, y, v)| \leq C(1 + |y| + |v|), \qquad (1.6)$$
$$|\Sigma(s, y, v)| \leq C(1 + |y| + |v|)$$

for all $s \in [t_0, t_1]$, $y \in I\!R$, and $v \in U$ with a constant $C > 0$. These conditions also imply that every control which has property (ii) is already admissible, i.e. it also meets (i) and (iii). This can be seen as follows: Suppose that conditions (1.5) and (1.6) are met and consider some control $u(\cdot)$ that has property (ii) of an admissible control. Using the notations of Section 1.1 let $A(t, y, \omega) = \Lambda(t, y, u(t, \omega))$ and $B(t, y, \omega) = \Sigma(t, y, u(t, \omega))$. Then the Lipschitz requirement of Theorem 1.1 follows from (1.5) by applying the mean value theorem and choosing $K \geq \max\{C, 1\}$. Without loss of generality we can assume that $2K^2 \geq 3C^2$ is valid. If we then choose $g(t) = \sqrt{1{,}5}C(1+|u(t)|)$ and $h(t) = C(1+u|(t)|)$, it follows from (1.6) that the growth condition of Theorem 1.1 is met. Hence, there exists a unique solution Y^u of the CSDE. Applying Theorem 1.2 one additionally obtains that $\mathrm{E}((\sup_{t\in[t_0,t_1]} |Y^u(t)|)^\rho) < \infty$ for $\rho \geq 2$ and therefore for all $\rho \in I\!N$. To this end, the conditions (1.5), (1.6), and property (ii) imply the admissibility of $u(\cdot)$.[8]

In our applications the requirements (1.5) and (1.6) are not always satisfied. But for another class of control problems at least condition (i) of an admissible control is valid if (ii) is satisfied. To obtain this result, the CSDE has to be one-dimensional and linear.

Corollary 1.1 (to the Variation of Constants) *Let $(t_0, y_0) \in Q$ and let $A_1^{(j)}$, $j = 1, \ldots d$, A_2, $B_1^{(i,j)}$, $i = 1, \ldots m$, $j = 1, \ldots d$, $B_2^{(i)}$, $i = 1, \ldots m$, be progressively measurable processes with*

[8] See Fleming/Soner (1993), p. 398.

$$\int_{t_0}^{t_1} |A_2(s)|\, ds \qquad < \infty \quad P\text{-}a.s.,\ t \geq 0,$$

$$\int_{t_0}^{t_1} \left(\sum_{j=1}^{d} A_1^{(j)}(s)^2 + \sum_{i=1}^{m} B_2^{(i)}(s)^2 \right) ds < \infty \quad P\text{-}a.s.,\ t \geq 0,$$

$$\int_{t_0}^{t_1} \left(\sum_{i=1}^{m} \sum_{j=1}^{d} B_1^{(i,j)}(s)^4 \right) ds \qquad < \infty \quad P\text{-}a.s.,\ t \geq 0.$$

Besides, let $u(\cdot)$ be a control with property (ii) of Definition 1.1. Then the linear CSDE

$$dY^u(t) = Y^u(t) \Big[(A_1(t)'u(t) + A_2(t))dt + (B_1(t)u(t) + B_2(t))'dW(t) \Big] \quad (1.7)$$

has the Lebesgue $\bigotimes$ P-unique solution

$$Y^u(t) = y_0 \cdot \exp\Bigg(\int_{t_0}^{t_1} \left(A_1(s)'u(s) + A_2(s) - 0,5|B_1(s)u(s) + B_2(s)|^2 \right) ds + \int_{t_0}^{t_1} \Big(B_1(s)u(s) + B_2(s) \Big)' dW(s) \Bigg).$$

In the case of bounded admissible controls the following integrability conditions are sufficient

$$\int_{t_0}^{t_1} \left(\sum_{j=1}^{d} |A_1^{(j)}(s)| + |A_2(s)| \right) ds \qquad < \infty \quad P\text{-a.s.},\ t \geq 0,$$

$$\int_{t_0}^{t_1} \left(\sum_{i=1}^{m} \sum_{j=1}^{d} B_1^{(i,j)}(s)^2 + \sum_{i=1}^{m} B_2^{(i)}(s)^2 \right) ds < \infty \quad P\text{-a.s.},\ t \geq 0.$$

Proof of Corollary 1.1. The assumed integrability requirements and property (ii) of an admissible control imply the integrability requirements of Proposition 1.1. □

In the following (1.7) is said to be a linear CSDE if the coefficients meet the integrability requirements of Corollary 1.1.

On the strength of the results on the existence and uniqueness of a solution of (1.1), we now turn to the question of which admissible control is the optimal one. This question is resolved by a so-called verification theorem.

In many applications the values of the state process do not lie in $\mathbb{R}^n$ but are restricted to an open subset. To cover this more general situation, let $\mathcal{O} \subset \mathbb{R}^n$ be an open subset of $\mathbb{R}^n$. In the case $\mathcal{O} \neq \mathbb{R}^n$ it is assumed that the boundary

$\partial\mathcal{O}$ is a compact, $(n-1)$-dimensional C^3-manifold. On the analogy of Q_0 we define $Q := [t_0, t_1) \times \mathcal{O}$. Since in the more general formulation of the problem it is not ensured that the state process never leaves the set $\mathcal{O}$, let

$$\tau := \inf\{t \in [t_0, t_1] : (t, Y(t)) \notin Q\}$$

be the first exit time of Y from $\mathcal{O}$. Hence,

$$(\tau, Y(\tau)) \in \partial^* Q := ([t_0, t_1) \times \partial\mathcal{O}) \cup (\{t_1\} \times \bar{\mathcal{O}}).$$

If $\mathcal{O} = I\!R^n$, we get $\partial^* Q = \{t_1\} \times I\!R^n$.

Suppose L and Ψ to be continuous, real-valued functions that for $k \in I\!N$ and a constant $C > 0$ meet the polynomial growth conditions

$$|L(t, y, v)| \leq C(1 + |y|^k + |v|^k), \tag{1.8}$$
$$|\Psi(t, y)| \leq C(1 + |y|^k) \tag{1.9}$$

on $\bar{Q} \times U$ and $\bar{Q}$, respectively. The functions L and Ψ measure the running and terminal utility resulting from the choice of a control. It is our goal to determine an admissible control for a fixed initial value (t_0, y_0) so that the utility functional

$$J(t_0, y_0; u) := \mathrm{E}^{t_0, y_0}\left(\int_{t_0}^{\tau} L(s, Y^u(s), u(s))\, dt + \Psi(\tau, Y^u(\tau))\right)$$

becomes maximal, i.e. $\max\limits_{u \in \mathcal{A}(t_0, y_0)} J(t_0, y_0; u)$.

We emphasize that the expected value involved in the utility functional is well-defined due to the polynomial growth conditions (1.8) and (1.9) as well as the properties (ii) and (iii) of an admissible control.

An important tool to solve a control problem is the value function which is defined by

$$V(t, y) := \sup_{u \in \mathcal{A}(t, y)} J(t, y; u), \quad (t, y) \in Q.$$

It measures the maximal utility given an initial condition $(t, y) \in Q$ and allows us to state a sufficient condition for a control to be optimal.[9] For this purpose, defining the differential operator

[9] The question of the existence of an optimal control is treated by Fleming/Rishel (1975, pp. 166ff) and Fleming/Soner (1993, pp. 167ff).

$$A^v G(t,y) := G_t(t,y) + 0.5 \sum_{i,j=1}^{n} \Sigma^*_{ij}(t,y,v) \cdot G_{y_i y_j}(t,y) + \sum_{i=1}^{n} \Lambda_i(t,y,v) \cdot G_{y_i}(t,y),$$

where $G \in C^{1,2}(Q)$, $(t,y) \in Q$, $v \in U$, and $\Sigma^* := \Sigma\Sigma'$, we obtain the following verification theorem.

Theorem 1.3 (Verification Theorem) *Suppose that the coefficients of the CSDE (1.1) satisfy the requirements (1.5) and (1.6). Besides, assume that the functions L and Ψ satisfy (1.8) and (1.9) and that there exists a function G with the following properties:*

a)

$$G \in C^{1,2}(Q) \cap C(\bar{Q}), \tag{1.10}$$

$$|G(t,y)| \leq K(1 + |y|^k), \tag{1.11}$$

with $K > 0$ and $k \in \mathbb{N}$,

b) G solves the Hamilton-Jacobi-Bellman equation (HJB):

$$\sup_{v \in U} \left\{ A^v G(t,y) + L(t,y,v) \right\} = 0, \qquad (t,y) \in Q, \tag{1.12}$$

$$G(t,y) = \Psi(t,y), \quad (t,y) \in \partial^* Q. \tag{1.13}$$

We then obtain the following results:

(i) $G(t,y) \geq J(t,y;u)$ for all $(t,y) \in Q$ and $u(\cdot) \in \mathcal{A}(t,y)$.

(ii) If for $(t,y) \in Q$ there exists a control $u^*(\cdot) \in \mathcal{A}(t,y)$ with

$$u^*(s) \in \arg\max_{v \in U} \left(A^v G(s, Y^*(s)) + L(s, Y^*(s), v) \right) \tag{1.14}$$

for all $s \in [t, \tau]$, where Y^* denotes the solution of the CSDE which belongs to $u^*(\cdot)$, then

$$G(t,y) = V(t,y) = J(t,y;u^*),$$

i.e. $u^*(\cdot)$ is an optimal control and G corresponds to the value function of the above problem.

A function G with properties (1.10) and (1.11) that solves the HJB (1.12) with boundary conditions (1.13) is said to be a *classical solution* with polynomial growth.

We wish to stress that in some important applications the requirements (1.5) and (1.6) as well as the growth condition (1.11) are not met. Below we will discuss how these requirements can be weakened.

Proof of Theorem 1.3.

(i) Let $(t,y) \in Q$ and let $u(\cdot) \in \mathcal{A}(t,y)$ be an admissible control. It is sufficient to show that for every stopping time θ with $t \le \theta \le \tau$ the inequality

$$G(t,y) \ge \mathrm{E}^{t,y}\left(\int_t^\theta L(s,Y(s),u(s))\,ds + G(\theta,Y(\theta))\right) \tag{1.15}$$

is valid.[10] Due to the boundary condition (1.13) of the HJB the claim (i) follows by choosing $\theta = \tau$.

First, we look at the case that the set $\mathcal{O}$ is *bounded.* Since G is supposed to be a solution of the HJB (1.12), we obtain

$$A^{u(s)}G(s,Y(s)) + L(s,Y(s),u(s)) \le 0 \tag{1.16}$$

for every admissible control $u(\cdot)$ and $s \in [t,\tau]$. Due to assumption a) the function G is a classical $C^{1,2}$-solution of the HJB and therefore Ito's formula can be applied to $G(\theta,Y(\theta))$:[11]

$$\begin{aligned} G(\theta,Y(\theta)) - G(t,y) - \int_t^\theta A^{u(s)}G(s,Y(s))\,ds \\ = \int_t^\theta G_y(s,Y(s)) \cdot \Sigma(s,Y(s),u(s))\,dW(s). \end{aligned} \tag{1.17}$$

With suitable constants K_i, $i = 1,\ldots,4$, we obtain for the integrand of the Ito integral

$$\begin{aligned} &\int_t^\theta |G_y(s,Y(s)) \cdot \Sigma(s,Y(s),u(s))|^2\,ds \\ &\qquad \le K_1 \int_t^\theta |\Sigma(s,Y(s),u(s))|^2\,ds \\ &\qquad \le K_2 \int_t^\theta (1 + |Y(s)| + |u(s)|)^2\,ds \\ &\qquad \le K_3 \int_t^\theta \left(|Y(s)|^2 + (1+|u(s)|)^2\right) ds \\ &\qquad \le K_4 \left(\mathrm{diam}(\mathcal{O})^2 + \int_t^\theta (1+|u(s)|)^2\,ds\right). \end{aligned}$$

[10] Note that, for notational convenience, we omit the index u of the state process $Y^u(t)$.

[11] Note that the solution Y^u belonging to the admissible control $u(\cdot)$ is a semi-martingale. See Protter (1990), pp. 193ff. For this reason Ito's formula can be applied. See Protter (1990, p. 74).

Here $\text{diam}(\mathcal{O})$ denotes the diameter of the set $\mathcal{O}$. The first estimate is valid because the continuous function G_y is bounded on the (bounded) set $[t, t_1] \times \mathcal{O}$. The second estimate follows from (1.6) and the third from $(v+w)^2 \leq 2v^2+2w^2$, $v, w \in I\!R$. The fourth estimate is obvious.

Since the set $\mathcal{O}$ is supposed to be bounded, we obtain $\text{diam}(\mathcal{O}) < \infty$. Due to property (ii) of an admissible control it follows

$$\mathrm{E}^{t,y}\left(\int_t^\theta |G_y(s, Y(s)) \cdot \Sigma(s, Y(s), u(s))|^2 \, ds\right) < \infty. \tag{1.18}$$

Hence, we get[12]

$$\mathrm{E}^{t,y}\left(\int_t^\theta G_y(s, Y(s)) \cdot \Sigma(s, Y(s), u(s)) \, dW(s)\right) = 0.$$

Taking expectations in (1.17) leads to

$$G(t, y) = \mathrm{E}^{t,y}\left(\int_t^\theta -A^{u(s)} G(s, Y(s)) \, ds + G(\theta, Y(\theta))\right). \tag{1.19}$$

Together with (1.16) this implies inequality (1.15).

Now let us consider the general case when the set $\mathcal{O}$ is *not* necessarily bounded. We prove relation (1.15) by approximating the set $\mathcal{O}$ via an increasing sequence of bounded sets

$$\mathcal{O}_p := \mathcal{O} \cap \{x \in I\!R^n : |x| < p, \ \text{dist}(x, \partial\mathcal{O}) > p^{-1}\}, \quad p \in I\!N.$$

For this purpose we define

$$Q_p := [t_0, t_1 - p^{-1}) \times \mathcal{O}_p,$$

where the sets Q_p are not empty for $p \in I\!N$ with $p > (t_1 - t_0)^{-1} =: \tilde{p}$. Without loss of generality we therefore assume $p > \tilde{p}$. Besides, let $\theta_p := \min\{\theta, \tau_p\}$ be a stopping time, where τ_p denotes the first exit time of $(s, Y(s))$ from Q_p. For $p \to \infty$ we get $\tau_p \to t_1$ P-a.s. and, consequently, $\theta_p \to \theta$ P-a.s.

Due to the boundedness of Q_p we again obtain

$$G(t, y) \geq \mathrm{E}^{t,y}\left(\int_t^{\theta_p} L(s, Y(s), u(s)) \, ds + G(\theta_p, Y(\theta_p))\right). \tag{1.20}$$

---

[12] See Gihman/Skorohod (1970), p. 29.

Hence, the claim (1.15) is valid if

$$\lim_{p\to\infty} \mathrm{E}^{t,y}\left(\int_t^{\theta_p} L(s,Y(s),u(s))\,ds\right) = \mathrm{E}^{t,y}\left(\int_t^{\theta} L(s,Y(s),u(s))\,ds\right), \tag{1.21}$$

$$\lim_{p\to\infty} \mathrm{E}^{t,y}\left(G(\theta_p,Y(\theta_p))\right) = \mathrm{E}^{t,y}\left(G(\theta,Y(\theta))\right). \tag{1.22}$$

Applying the theorem of dominated convergence, property (1.21) is proved if

$$\left\{\int_t^{\theta_p} L(s,Y(s),u(s))\,ds\right\}_p \text{ is uniformly integrable and} \tag{1.23}$$

$$\lim_{p\to\infty}\int_t^{\theta_p} L(s,Y(s),u(s))\,ds = \int_t^{\theta} L(s,Y(s),u(s))\,ds \quad P\text{-a.s.} \tag{1.24}$$

and property (1.22) is proved if

$$\{G(\theta_p,Y(\theta_p))\}_p \text{ is uniformly integrable and} \tag{1.25}$$

$$\lim_{p\to\infty} G(\theta_p,Y(\theta_p)) = G(\theta,Y(\theta)) \quad P\text{-a.s..} \tag{1.26}$$

If these properties are proved, assertion (i) of Theorem 1.3 is established. Recall that a sequence of random variables $(X_p)_{p\in\mathbb{N}}$ is uniformly integrable if there exists $\rho > 1$ such that[13]

$$\sup_{p\in\mathbb{N}} \mathrm{E}(|X_p|^\rho) < \infty. \tag{1.27}$$

Proof of (1.23). Consider the following estimate

$$\begin{aligned}
&\left(\int_t^{\theta_p} L(s,Y(s),u(s))\,ds\right)^2\\
&\quad\le \left(\int_t^{\theta_p} |L(s,Y(s),u(s))|\,ds\right)^2\\
&\quad\le C^2\left(\int_t^{t_1}\left(1+|u(s)|^k+|Y(s)|^k\right)ds\right)^2\\
&\quad\le C^2(t_1-t)\int_t^{t_1}\left(1+|u(s)|^k+|Y(s)|^k\right)^2 ds\\
&\quad\le 3C^2(t_1-t)\int_t^{t_1}\left(1+|u(s)|^{2k}+|Y(s)|^{2k}\right)ds\\
&\quad\le 3C^2(t_1-t)\cdot\left[\int_t^{t_1}|u(s)|^{2k}\,ds + (t_1-t)\cdot\left(1+\sup_{s\in[t,t_1]}|Y(s)|^{2k}\right)\right].
\end{aligned}$$

[13] See Billingsley (1986), pp. 220f.

The first estimate follows from the triangle inequality of Lebesgue integrals. The second is valid due to the polynomial growth condition (1.8) on L as well as the positivity of the integrand and $\theta_p \leq t_1$. The third estimate follows from the Hölder inequality and the fourth from $(v_1+v_2+v_3)^2 \leq 3v_1^2+3v_2^2+3v_3^2$, $v_1, v_2, v_3 \in I\!R$.

Due to the properties (ii) and (iii) of an admissible control we conclude

$$\sup_{p>\tilde{p}} \mathrm{E}^{t,y} \left(\int_t^{\theta_p} L(s,Y(s),u(s))\, ds \right)^2 < \infty,$$

which implies the uniform integrability of (1.23).

Proof of (1.24). Applying again the theorem of dominated convergence, (1.24) follows. Note that the sequence of functions $(1_{[t,\theta_p]} L(s,Y(s),u(s)))_p$ can be dominated because of the polynomial growth condition (1.8) as well as properties (ii) and (iii) of an admissible control.

Proof of (1.25). Due to the growth condition (1.11) on G we obtain the following estimate for fixed $k \in I\!N$

$$\begin{aligned} (G(\theta_p, Y(\theta_p)))^2 &\leq K^2(1+|Y(\theta_p)|^k)^2 \\ &\leq 2K^2(1+|Y(\theta_p)|^{2k}) \\ &\leq 2K^2(1+\sup_{s\in[t,t_1]} |Y(s)|^{2k}). \end{aligned}$$

Hence, from property (iii) of an admissible control we conclude that

$$\sup_{p>\tilde{p}} \mathrm{E}^{t,y} (G(\theta_p, Y(\theta_p)))^2 < \infty,$$

which implies the uniform integrability of the family $\{G(\theta_p, Y(\theta_p))\}_p$.

Proof of (1.26). Due to the continuity of G we obtain (1.26).

(ii) For fixed $(t,y) \in Q$ and an admissible control $u^*(\cdot)$ with (1.14) it follows

$$A^{u^*(s)} G(s, Y^*(s)) + L(s, Y^*(s), u^*(s)) = 0$$

for all $s \in [t,\tau]$. Here Y^* denotes the state process belonging to $u^*(\cdot)$. With (1.19) we obtain for the stopping time θ_p

$$G(t,y) = \mathrm{E}^{t,y} \left(\int_t^{\theta_p} L(s, Y^*(s), u^*(s))\, ds + G(\theta_p, Y^*(\theta_p)) \right).$$

Taking the limit $p \to \infty$, by (1.21) and (1.22) we get

$$G(t,y) = \mathrm{E}^{t,y}\left(\int_t^\theta L(s, Y^*(s), u^*(s))\, ds + G(\theta, Y^*(\theta))\right).$$

Choosing $\theta = \tau$ it follows $G(t,y) = J(t,y;u^*)$. Combining this result with part (i) of the verification theorem we get

$$J(t,y;u^*) = G(t,y) \overset{\text{(i)}}{\geq} \underbrace{\sup_{u \in \mathcal{A}(t,y)} J(t,y;u)}_{=V(t,y)} \geq J(t,y;u^*).$$

This establishes (ii). □

Recalling the results of the verification theorem and its proof, we arrive at the following facts:

- Conditions (1.5) and (1.6) ensure the existence and uniqueness of a solution of the CSDE for controls that meet property (ii) of Definition 1.1. Ito's formula can be applied to such a solution and therefore condition (1.17) follows.
- Growth condition (1.6) is applied to prove condition (1.18).
- Growth condition (1.11) is applied to prove condition (1.25).

In the proof of the following corollary we show that the results of the verification theorem are still valid for linear CSDEs if (1.25) can be verified directly. In this case, conditions (1.5), (1.6), and (1.11) are not required.

Corollary 1.2 (to the Verification Theorem) *Consider a linear CSDE with coefficients satisfying the requirements of Corollary 1.1 and*[14]

$$\mathrm{E}\left(\int_{t_0}^{t_1}\left\{\sum_{i=1}^m B_2^{(i)}(s)^2 + \sum_{i=1}^m\sum_{j=1}^d B_1^{(i,j)}(s)^4\right\} ds\right) < \infty, \qquad t \geq 0. \qquad (1.28)$$

Suppose that the requirements (1.8) and (1.9) on the functions L and ψ hold. Besides, assume that there exists a function $G \in C^{1,2}(Q) \cap C(\bar{Q})$ solving the HJB (1.12) with boundary condition (1.13). For all $(t,y) \in Q$ and all admissible controls $u(\cdot) \in \mathcal{A}(t,y)$ it is assumed that the family

[14] The following condition is not explicitly mentioned in Korn/Kraft (2001), but it is of course satisfied in all examples treated there.

$$\{G(\theta_p, Y(\theta_p))\}_p \textit{ is uniformly integrable,} \tag{1.25}$$

where θ_p is defined as in the proof of the verification theorem. Then the results (i) and (ii) of the verification theorem are still valid.

For $\rho > 1$, for all $(t, y) \in Q$, for all admissible controls $u(\cdot) \in \mathcal{A}(t, y)$, and every stopping time θ with $t \le \theta \le t_1$ we obtain

$$|G(\theta, Y(\theta))|^\rho \le \sup_{s\in[t,t_1]} |G(s, Y(s))|^\rho$$

and thus

$$\mathrm{E}(|G(\theta, Y(\theta))|^\rho) \le \mathrm{E}\Big(\sup_{s\in[t,t_1]} |G(s, Y(s))|^\rho \Big).$$

Hence, due to (1.27) it is sufficient for (1.25) if for all $(t, y) \in [t_0, t_1] \times \mathcal{O}$ and all admissible controls $u(\cdot) \in \mathcal{A}(t, y)$ there exists a constant $\rho > 1$ such that

$$\mathrm{E}\Big(\sup_{s\in[t,t_1]} |G(s, Y(s))|^\rho \Big) < \infty. \tag{1.29}$$

Proof of Corollary 1.2.
For admissible controls the linear CSDE possesses a unique solution which is explicitly given in Corollary 1.1. Ito's formula can be applied to such a solution. Hence, relation (1.17) remains valid.

To prove property (1.18), recall that the diffusion coefficient of a linear CSDE reads $\Sigma(t, y, v) = y(B_1(t)v + B_2(t))$. In analogy to the proof of the verification theorem we assume that the set $\mathcal{O}$ is bounded. For any admissible control $u(\cdot)$ we thus obtain the estimate

$$\begin{aligned}
\int_{t_0}^{t_1} |\Sigma(s, Y(s), u(s))|^2\, ds &= \int_{t_0}^{t_1} |Y(s)(B_1(s)u(s) + B_2(s))|^2\, ds \\
&\le \sup_{s\in[t,t_1]} |Y(s)|^2 \int_{t_0}^{t_1} (|B_1(s)u(s)| + |B_2(s)|)^2\, ds \\
&\le 2\,\mathrm{diam}(\mathcal{O}) \int_{t_0}^{t_1} |B_1(s)u(s)|^2 + |B_2(s)|^2\, ds \\
&\le 2\,\mathrm{diam}(\mathcal{O}) \int_{t_0}^{t_1} |B_1(s)|^4 + |u(s)|^4 + |B_2(s)|^2\, ds.
\end{aligned}$$

The last inequality is valid due to $|Mv| \le |M||v|$, $M \in \mathbb{R}^{m,d}$, $v \in \mathbb{R}^d$, and $2w_1w_2 \le w_1^2 + w_2^2$, w_1, $w_2 \in \mathbb{R}$. Then (1.18) follows from property (ii) of an admissible control and from requirement (1.28). Note that the operator

norm $|\cdot|$ and the row-sum norm $|\cdot|_\infty$ are equivalent and that $|B_1(s)|_\infty \le \sum_{i=1}^m \sum_{j=1}^d |B_1^{(i,j)}(s)|$.
Hence, given a linear CSDE one can abandon conditions (1.5) and (1.6) and the results of the verification theorem can be established if requirement (1.25) is satisfied. □

The verification theorem and the corresponding corollary lead to the following three step algorithm:[15]

1st step: For fixed $(t, y) \in Q$ one solves the static optimization problem

$$\max_{v \in U} \left(A^v G(t,y) + L(t,y,v) \right),$$

where G is formally assumed to be given. We obtain an optimal value v^* which depends on t, y, G, and the corresponding partial derivatives of G. Interpreting this result as a function of t and y, we get a function $u^*(t, y; G)$.

2nd step: The function $u^*(t, y; G)$ is the candidate for the optimal control, where the function G still has to be determined. For this purpose, $u^*(t, y; G)$ is plugged into the HJB (1.12). Since $u^*(t, y; G)$ solves the optimization problem of the first step, one can drop the supremum to get a PDE for G:

$$A^{u^*(t,y;G)} G(t,y) + L(t,y,u^*(t,y;G)) = 0, \qquad (t,y) \in Q, \tag{1.30}$$
$$G(t,y) = \Psi(t,y),\ (t,y) \in \partial^* Q.$$

Substituting the solution of (1.30) and the state process Y^* into the function $u^*(t, y; G)$ leads to a candidate $u^*(\cdot)$ for the optimal control that does not involve the function G. Here Y^* is the state process for the control $u^*(\cdot)$.

3rd step: Next all assumptions have to be verified that are necessary to apply the verification theorem or Corollary 1.2. In particular, one needs to show that $u^*(\cdot)$ is admissible and that G has all properties of a value function. Table 1.1 summarizes those conditions which have to be verified.

[15] See Björk (1998, p. 209) and Korn/Korn (2001, p. 232).

Table 1.1: Assumptions of the verification theorem and Corollary 1.2

	Verification Theorem	Corollary 1.2
Main requirements	Growth condition (1.5) and (1.6) on Λ and Σ	Linear CSDE satisfying (1.28)
Conditions on $u^*(\cdot)$	Progressively measurability	
	Property (ii) of an admissible control	
		Property (iii) of an admissible control
Conditions on G	$C^{1,2}$-solution of the HJB	
	Growth condition (1.11)	(1.25) or (1.29)

It is worth pointing out that one can get rid of the requirements (1.8) and (1.9) on L and Ψ. As stated above, (1.8) and (1.9) are required to ensure that the value function is well-defined. Apart from this, (1.8) is applied to prove the verification theorem, but both (1.8) and (1.9) are not applied to prove Corollary 1.2. Hence, Corollary 1.2 is still valid if we does not require (1.8) and (1.9), but instead show that for each admissible control u and each initial value (t_0, y_0) the utility functional $J(t_0, y_0; u)$ is well-defined.

Instead of making the assumption of Corollary 1.2 that all admissible controls meet (1.25), one can proceed in the following way: Since in our applications we look at portfolio problems without intermediate consumption, we assume for simplicity that $L \equiv 0$. Then, given a candidate G for the value function one defines a set of controls with the following properties:

(i) u is progressively measurable,

(ii) for all initial conditions $y_0 > 0$ the corresponding state process Y^u with $Y(0) = y_0$ has a pathwise unique solution $\{Y^u(t)\}_{t \in [t_0, t_1]}$,

(iii) $\mathrm{E}\Big(\int_{t_0}^{t_1} |u(s)|^4 \, ds \Big) < \infty$,

(iv) $\{G(\theta_p, Y^u(\theta_p))\}$ meets (1.25),

(v) the utility functional $J(t_0, y_0; u)$ is well-defined.

This set is denoted by $\tilde{\mathcal{A}}(t_0, y_0)$ and the elements are said to be *weakly admissible.* If the candidate for the optimal control u^* corresponding to G is an element of $\tilde{\mathcal{A}}(0, x_0)$, the process u^* is an optimal control among all weakly admissible controls. This result is stated in the following corollary:

Corollary 1.3 (Variant of Corollary 1.2) *Assume $L \equiv 0$ and consider a linear CSDE whose coefficients meet (1.28) and the requirements of Corollary 1.1. Besides, assume that there exists a function $G \in C^{1,2}(Q) \cap C(\bar{Q})$ that solves the HJB (1.12) with boundary condition (1.13). Further, suppose that for $(t, y) \in Q$ there exists a weakly admissible control $u^*(\cdot) \in \tilde{\mathcal{A}}(t, y)$ with*

$$u^*(s) \in \arg\max_{v \in U} A^v G(s, Y^*(s))$$

for all $s \in [t, \tau]$, where Y^* denotes the solution of the CSDE which belongs to $u^*(\cdot)$. Then the following results are valid:

(i) $G(t, y) \geq J(t, y; u)$ for all $(t, y) \in Q$ and $u(\cdot) \in \tilde{\mathcal{A}}(t, y)$.

(ii) Besides, $u^*(\cdot)$ is an optimal control among all weakly admissible controls and G corresponds to the value function of the optimization problem over all weakly admissible controls.

Finally, let us stress that the assumption $L \equiv 0$ is not crucial. However, some requirement is needed to ensure (1.23).

2

Optimal Portfolios with Stochastic Interest Rates

2.1 Introduction

The continuous-time portfolio problem has its origin in the pioneering work of Merton (1969, 1971), which deals with finding the optimal investment strategy of an investor. More precisely, the investor looks for an optimal decision on how many shares of which security he should hold at any time between now and a time horizon T to maximize expected utility from terminal wealth. In the classical Merton problem the investor can allocate money to a riskless money market account (syn. savings account) and to d different stocks of varying degrees of risk. By describing the actions of the investor via the portfolio process (i.e. the percentages of wealth invested in the different securities) Merton was able to reduce the portfolio problem to a control problem which could be solved by using standard stochastic control methodology.[1] A drawback of Merton's model, however, is the assumption of deterministic interest rates. Our main objective in the current chapter is to overcome this restriction. Therefore, we analyze portfolio problems in which the interest rate dynamics of the economy can be described by the Vasicek model, the Dothan model, the Black-Karasinski model, or the Cox-Ingersoll-Ross model. Our presentation of the problem with Ho-Lee or Vasicek term structure goes back to Korn/Kraft (2001), who applied stochastic control theory to arrive at the solution.[2] By contrast, the results concerning the Dothan and Black-Karasinski models are

[1] The other main approach to solve portfolio problems is the martingale method. We refer to Korn (1997) for an introduction to this method.

[2] Sørensen (1999) applied the martingale approach to solve a related problem.

new. The portfolio problem with Cox-Ingersoll-Ross term structure was partly solved by Deelstra/Graselli/Koehl (2000) applying the martingale approach under the assumption of a complete market. In contrast to their approach we apply stochastic control theory so that we can drop the assumption of completeness. Besides, we are able to solve the problem for a set of market coefficients not covered by the results of Deelstra/Graselli/Koehl (2000).

On the theoretical side, the introduction of stochastic interest rates into the portfolio problem means that the stochastic differential equation describing the wealth process does not satisfy the usual Lipschitz assumptions needed to apply standard verification theorems. However, due to the special structure of this equation, which is called the wealth equation, a suitable verification result of Chapter 1 can be applied.

2.2 Ho-Lee and Vasicek Model

We consider an economy with $d+1$ assets, which are continuously traded on a frictionless market. All traders are assumed to be price takers. The uncertainty is modeled by a probability space $(\Omega, \mathcal{F}, P)$. On this space an m-dimensional Brownian motion $\{(W(t), \mathcal{F}_t)\}_{t\geq 0}$ is defined where $\{\mathcal{F}_t\}_{t\geq 0}$ denotes the Brownian filtration. One of the assets is a savings account (syn. money market account) following the differential equation

$$dM(t) = M(t)r(t)dt$$

with $M(0) = 1$. Here r denotes the short rate, which can be interpreted as the annualized interest for the infinitesimal period $[t, t+dt]$.

In contrast to Merton's classical model,[3] we assume a short rate modeled by the SDE

$$dr(t) = a(t)dt + b\,dW_1(t),$$

$t \in [0, T^*]$, $b > 0$, with initial data $r(0) = r_0$. In this section we consider the Ho-Lee model given by $a(t) = \tilde{a}(t) + b\zeta(t)$ and the Vasicek approach with $a(t) = \theta(t) - \alpha r(t) + b\zeta(t)$, $\alpha > 0$, respectively. The market price of risk

[3] See Merton (1969, 1971, 1990), Fleming/Rishel (1975), pp. 160f, Duffie (1992), pp.145ff, Fleming/Soner (1993), pp. 174ff, Korn (1997), pp. 48ff.

ζ is assumed to be deterministic and continuous which implies the progressive measurability of ζ. In particular, this assumption guarantees that ζ is bounded on each compact interval. Furthermore, let the initial forward rate curve $f^*(0,T)$, $0 \le T \le T^*$, be continuously differentiable, which leads to $\tilde{a}(t) = f_T^*(0,t) + b^2 t$ and $\theta(t) = f_T^*(0,t) + \alpha f^*(0,t) + \frac{b^2}{2\alpha}(1 - e^{-2\alpha t})$.[4] The price processes of the remaining d assets, which can be stocks and/or (zero) bonds, are assumed to follow Ito processes of the form

$$dP_i(t) = P_i(t)\Big[\mu_i(t)dt + \sigma_i(t)dW(t)\Big]$$

with $P_i(0) = p_i > 0$ and where $\mu(\cdot)$ is $I\!R^d$-valued and $\sigma_i(\cdot)$ denotes the i th row of the $d \times m$-matrix $\sigma(\cdot)$.

We consider an investor who starts with an initial wealth of $x_0 > 0$ at time $t = 0$. In the beginning this initial wealth is invested in the different assets and he is allowed to adjust his holdings continuously up to a fixed planning horizon T. His investment behavior is modeled by a portfolio process $\pi = (\pi_1, \ldots, \pi_d)$ which is progressively measurable (with respect to $\{\mathcal{F}_t\}_{t \ge 0}$). Here, $\pi_i(t)$, $i = 0, \ldots, d$, denotes the percentage of total wealth invested in the i-th asset at time t. Obviously, the percentage invested in the savings account is given by $1 - \pi'\underline{1}$ where $\underline{1} := (1, \ldots, 1)' \in I\!R^d$.

If we restrict our considerations to self-financing portfolio processes, his wealth process follows the stochastic differential equation (SDE)

$$dX(t) = X(t)\Big[\big(\pi(t)'(\mu(t) - r(t) \cdot \underline{1}) + r(t)\big)dt + \pi(t)'\sigma(t)dW(t)\Big] \tag{2.1}$$

with $X(0) = x_0$.[5]

The wealth equation can be interpreted as a controlled SDE with the control being the portfolio process $\pi(\cdot)$. In this setting the investor chooses a portfolio process to maximize utility. We assume that his preferences can be represented by the utility function $U(x) = x^\gamma$, $x \ge 0$, $0 < \gamma < 1$. Furthermore, the investor is only allowed to pick out a portfolio process which is admissible in the sense of Definition 1.1 and leads to a *positive* wealth process X^π. We are now in a position to formulate his optimization problem:[6]

[4] See for example Musiela/Rutkowski (1997), pp. 323f.

[5] See for example Korn/Korn (2001), pp. 62f.

[6] Here $\mathcal{A}(0, x_0)$ denotes the set of all admissible controls corresponding to the initial condition $(0, x_0)$. See Definition 1.1 in the Chapter 1.

$$\max_{\pi(\cdot)\in\mathcal{A}^*(0,x_0)} \mathrm{E}(X^\pi(T))^\gamma$$

with

$$dX^\pi(t) = X^\pi(t)\Big[(\pi(t)'(\mu(t) - r(t)\cdot \underline{1}) + r(t))dt + \pi(t)\sigma(t)dW(t)\Big],$$
$$X^\pi(0) = x_0$$

and

$$\mathcal{A}^*(0,x_0) := \Big\{\pi(\cdot)\in\mathcal{A}(0,x_0)\ :\ X^\pi(s)\geq 0\ P-\text{f.s. for } s\in[0,T]\Big\}.$$

We emphasize that applying optimal control methods to this problem does not automatically yield a positive state process. However, Corollary 1.1 and the specific form of the coefficients in the wealth equation (2.1) guarantee indeed the positivity of $X^\pi(t)$. Therefore, we obtain $\mathcal{A}^*(0,x_0) = \mathcal{A}(0,x_0)$.

2.2.1 Bond Portfolio Problem

We start in considering a portfolio problem where the investor can split up wealth in a savings account and a (zero) bond with maturity $T_1 > T$.[7] We assume that the asset price processes can be represented by the Ito processes

$$dM(t) = M(t)r(t)dt,$$
$$dP(t,T_1) = P(t,T_1)\Big[(\underbrace{r(t)+\zeta(t)\sigma(t)}_{=:\mu(t)})dt + \sigma(t)dW(t)\Big],$$

where W is a one-dimensional Brownian motion. In the Ho-Lee and the Vasicek models the volatility of the bond is given by $\sigma(t) = -b(T_1-t)$ and $\sigma(t) = \frac{b}{a}(\exp(-a(T_1-t))-1)$, respectively.[8] Let $\pi(t)$ be the percentage invested in the bond. This leads to a wealth equation of the form

$$\begin{aligned} dX(t) &= X(t)\Big[(\pi(t)\mu(t) + (1-\pi(t))r(t))dt + \pi(t)\sigma(t)dW(t)\Big] \quad (2.2)\\ &= X(t)\Big[(\pi(t)\zeta(t)\sigma(t) + r(t))dt + \pi(t)\sigma(t)dW(t)\Big] \end{aligned}$$

with initial wealth $X(0) = x_0$.

[7] In Chapter 3 we will see that, by applying the elasticity approach to portfolio optimization, we can also deal with the case $T_1 \leq T$.

[8] See for example Musiela/Rutkowski (1997), pp. 323ff.

As in contrast to the classical Merton problem, we assume a stochastic short rate, the drift coefficient includes the additional stochastic term $r(t)$. Thus, to solve the portfolio problem (2) by stochastic control methods we need to look at a two-dimensional state process $Y = (X, r)$. Note that the second component cannot be controlled via $\pi(\cdot)$. Using the notation of (1.1) in Chapter 1 we get[9]

$$\begin{aligned}
Y(t) &= (X(t), r(t))', \\
\Lambda(t, x, r, \pi) &= (x(\pi\zeta\sigma + r), a)', \\
\Sigma(t, x, r, \pi) &= (x\pi\sigma, b)', \\
\Sigma^*(t, x, r, \pi) &= \begin{pmatrix} x^2\pi^2\sigma^2 & bx\pi\sigma \\ bx\pi\sigma & b^2 \end{pmatrix}, \\
A^\pi G(t, x, r) &= G_t + 0.5(x^2\pi^2\sigma^2 G_{xx} + 2x\pi b\sigma G_{xr} + b^2 G_{rr}) \\
&\quad + x(\pi\zeta\sigma + r)G_x + aG_r.
\end{aligned}$$

Hence, the following Hamilton-Jacobi-Bellman equation (HJB) needs to be solved

$$\begin{aligned}
\sup_{|\pi| \le \delta} A^\pi G(t, x, r) &= 0, \\
G(T, x, r) &= x^\gamma,
\end{aligned}$$

where $\delta > 0$ will be specified later.

Note that due to the presence of the product rx in the above setting usual verification theorems which require Lipschitz conditions are not applicable to our situation as both the wealth process and the short rate are unbounded processes. We therefore apply Corollary 1.2. This result allows us to solve the HJB with the usual three step procedure. We would like to underline that in our opinion the third step, verification of all assumptions of both Corollary 1.2 and those made to perform the following calculations, forms an essential part of the solution.

Let us start with the calculation of the optimal bond position $\pi(\cdot)$.

1st step: Assuming $G_{xx} < 0$ we obtain the following candidate for the optimal bond position

[9] For the sake of simplicity we omit at times the functional dependencies with respect to t, x and r.

$$\pi^* = -\frac{\zeta}{\sigma}\frac{G_x}{xG_{xx}} - \frac{b}{\sigma}\frac{G_{xr}}{xG_{xx}}. \tag{2.3}$$

2nd step: Substituting $\pi^*(t, x, r; G)$ into the HJB leads to the PDE

$$\begin{aligned} 0 = {} & G_t G_{xx} - 0.5\zeta^2 G_x^2 - 0.5b^2 G_{xr}^2 + 0.5b^2 G_{rr} G_{xx} \\ & -b\zeta G_x G_{xr} + xrG_x G_{xx} + aG_r G_{xx} \end{aligned} \tag{2.4}$$

with terminal condition $G(T, x, r) = x^\gamma$. Note that $\zeta = (\mu - r)/\sigma$.

The form of this condition recommends the following separation ansatz

$$G(t, x, r) = f(t, r) \cdot x^\gamma \quad \text{with } f(T, r) = 1 \text{ for all } r.$$

This leads to a second-order PDE for f of the form

$$\begin{aligned} 0 = {} & (\gamma - 1) f f_t - 0.5b^2 \gamma f_r^2 - 0.5\zeta^2 \gamma f^2 + 0.5b^2(\gamma - 1) f f_{rr} \\ & -b\zeta\gamma f f_r + r\gamma(\gamma - 1) f^2 + a(\gamma - 1) f f_r \end{aligned}$$

with terminal condition $f(T, r) = 1$. Using the ansatz

$$f(t, r) = g(t) \cdot \exp(\beta(t) \cdot r)$$

with terminal conditions $\beta(T) = 0$ and $g(T) = 1$ and simplification yields

$$\begin{aligned} 0 = {} & (\gamma - 1) \cdot g' + (\gamma - 1)\left(\gamma + \beta'\right) \cdot rg \\ & - \left(0.5\zeta^2\gamma + 0.5b^2\beta^2 + b\zeta\gamma\beta\right) \cdot g + a(\gamma - 1)\beta \cdot g. \end{aligned} \tag{2.5}$$

Our ansatz for f will only be meaningful if we arrive at an ordinary differential equation (ODE) for g which does not include the short rate r.

In the Ho-Lee model the drift a of the short rate is a function of t, whereas in the Vasicek model it is a function of t and r. Therefore, we treat the two interest rate models separately.

Ho-Lee Model

In our Ho-Lee setting PDE (2.5) has the form

$$\begin{aligned} 0 = {} & (\gamma - 1) \cdot g' + (\gamma - 1)\left(\gamma + \beta'\right) \cdot rg \\ & + \underbrace{\left(-0.5\zeta^2\gamma - 0.5b^2\beta^2 - b\zeta\gamma\beta + a(\gamma - 1)\beta\right)}_{=:h_1(t)} \cdot g. \end{aligned} \tag{2.6}$$

Since $a(t) = f_T^*(0,t) + b^2 t + b\zeta(t)$ and ζ is assumed to be deterministic as well as continuous, h_1 is a continuous and deterministic function. Choosing $\beta(t) = \gamma(T-t)$ we obtain from (2.6) the following first-order ODE for g:

$$0 = (\gamma - 1) \cdot g' + h_1(t) \cdot g$$

with $g(T) = 1$. Separation of variables leads to

$$g(t) = \exp\left(\tfrac{1}{1-\gamma}(H_1(t) - H_1(T))\right),$$

where H_1 is a primitive of h_1. Hence, we obtain

$$G(t,x,r) = x^\gamma \cdot \exp\left(\tfrac{1}{1-\gamma}(H_1(t) - H_1(T)) + \gamma(T-t)r\right)$$

as a candidate for the value function. Substituting this into (2.3) gives the corresponding control

$$\begin{aligned}\pi^*(t) &= \frac{1}{1-\gamma} \cdot \frac{\zeta(t) + b\beta(t)}{-\sigma(t)} \\ &= \frac{1}{1-\gamma} \cdot \frac{\zeta(t) + b(T-t)\gamma}{-b(T_1 - t)}.\end{aligned}$$

Obviously, $\pi^*(\cdot)$ is continuous, deterministic and therefore bounded.

Vasicek Model

With Vasicek's specification of a the PDE (2.5) has the following form

$$\begin{aligned} 0 = (\gamma - 1) \cdot g' + &\underbrace{(\gamma - 1)(\beta' - \alpha\beta + \gamma)}_{(*)} \cdot rg \\ &+ \underbrace{(\theta(\gamma - 1)\beta - b\zeta\beta - 0.5b^2\beta^2 - 0.5\zeta^2\gamma)}_{=:h_2(t)} \cdot g. \end{aligned}$$

Our ansatz for f is only meaningful if β can be calculated so that the factor $(*)$ becomes zero. As a result, we need to solve an inhomogeneous ODE for β which has the following form

$$\beta'(t) = \alpha\beta(t) - \gamma$$

with $\beta(T) = 0$ leading to

$$\beta(t) = \tfrac{\gamma}{\alpha}(1 - \exp(\alpha(t - T))).$$

Choosing β as calculated, we again get a first-order homogeneous ODE for g

$$0 = (\gamma - 1) \cdot g' + h_2(t) \cdot g$$

with $g(T) = 1$. Hence

$$g(t) = \exp(\tfrac{1}{1-\gamma}(H_2(t) - H_2(T))),$$

where H_2 is a primitive of h_2. Therefore

$$G(t,x,r) = x^\gamma \cdot \exp\left(\tfrac{1}{1-\gamma}(H_2(t) - H_2(T)) + \tfrac{\gamma}{\alpha}(1 - \exp(\alpha(t-T)))r\right).$$

The corresponding control reads as follows

$$\begin{aligned}\pi^*(t) &= \frac{1}{1-\gamma} \cdot \frac{\zeta(t) + b\beta(t)}{\sigma(t)} \\ &= \frac{1}{1-\gamma} \cdot \frac{\zeta(t) + b \cdot \frac{\gamma}{\alpha}(1 - \exp(\alpha(t-T)))}{\frac{b}{\alpha}(\exp(-\alpha(T_1 - t)) - 1)}.\end{aligned}$$

Again $\pi^*(\cdot)$ is continuous, deterministic and therefore bounded.

In both cases δ can be so chosen that the optimal bond position fulfils the condition $|\pi(\cdot)| \leq \delta$. Moreover, the respective $\pi^*(\cdot)$ takes the form

$$\pi^*(t) = \underbrace{\frac{1}{1-\gamma} \cdot \frac{\zeta(t)}{\sigma(t)}}_{\text{Merton result}} - \underbrace{\frac{\gamma}{1-\gamma} \cdot \kappa(t)}_{\text{correction term}}$$

with $\kappa(t) = \frac{T-t}{T_1-t}$ in the Ho-Lee model and $\kappa(t) = \frac{1-e^{-\alpha(T-t)}}{1-e^{-\alpha(T_1-t)}}$ in the Vasicek model. The first term coincides with the classical optimal solution of Merton (1969, 1971) when the coefficients are deterministic. The second term can be interpreted as a correction term which is positive and monotonously decreasing to zero up to the terminal date T. Thus, we first have a bigger, negative deviation from the classical result which disappears at the time horizon. This correction comes from the fact that the investor tries to hedge his portfolio against interest rate risk. We will provide further interpretations of these results in Chapter 3.

3rd step: At first we justify our use of Corollary 1.2, although the state process $Y = (X, r)'$ is two-dimensional: Note that the short rate process does not include the control $\pi(\cdot)$. Therefore one can prove condition (i) and (iii) in Definition 1.1 independently of a specified control. Consider the SDE

$$dr(t) = a(t)dt + b\,dW(t) \tag{2.7}$$

of the short rate r with $r(0) = r_0$. The coefficients meet the growth and Lipschitz conditions of the existence and uniqueness theorem for SDE.[10] Hence, (2.7) has a unique solution. Using a theorem of Krylov (1980, p. 85), we arrive at

$$\mathrm{E}\Big(\max_{0\le s\le T} |r(s)|^\rho\Big) < +\infty \tag{2.8}$$

with $\rho \in I\!N$. Therefore, independently of the control under consideration, the conditions (i) and (iii) are fulfilled by the second component of the state process Y. As a result we can treat our problem as if the state process consists only of X. Note that under this condition the wealth equation is a linear controlled SDE.

We can apply Corollary 1.2 if we are able to prove the following assumptions:

1) $\pi^*(\cdot)$ is progressively measurable,
2) $\pi^*(\cdot)$ meets condition (ii) of Definition 1.1,
3) $\pi^*(\cdot)$ meets condition (iii) of Definition 1.1,
4) G is a $C^{1,2}$-solution of the HJB,
5) condition (1.29) is met,

Further, the portfolio process has to lead to a positive wealth process, i.e.

6) $X^{\pi^*} \ge 0$.

Proof of 1). The respective solution $\pi^*(\cdot)$ is continuous and deterministic. Therefore, it is progressively measurable.

Proof of 2). Property (ii) of an admissible control is met because the respective $\pi^*(\cdot)$ is bounded.

Proof of 3). By Corollary 1.1 the wealth equation (2.2) for $\pi^*(\cdot)$ has the solution

$$X^*(t) = x_0 \exp\bigg(\int_0^t \pi^*(s)\zeta(s)\sigma(s) + r(s) - 0.5(\pi^*(s)\sigma(s))^2\, ds + \int_0^t \pi^*(s)\sigma(s)\, dW(s)\bigg). \tag{2.9}$$

[10] See Fleming/Soner (1993, pp. 397f).

Note that (2.8) implies

$$\mathrm{E}\left(\left|\int_0^T r(s)\,ds\right|\right) \leq T \cdot \mathrm{E}\Big(\max_{0\leq s\leq T} |r(s)|\Big) < +\infty$$

and thus

$$\int_0^T r(s)\,ds < +\infty, \quad P-\text{f.s.}.$$

Obviously the other assumptions of Corollary 1.1 are met.

With an appropriate constant $K > 0$ we obtain the following estimate. (Note that $\pi^*(\cdot)$, $\sigma(\cdot)$, and $\zeta(\cdot)$ are bounded and that $|uv| \leq u^2 + v^2$ for $u, v \in I\!R$.):

$$\begin{aligned} X^*(t)^k &= x_0^k \cdot \exp\left(k\int_0^t \pi^*(s)\zeta(s)\sigma(s) + r(s) - 0.5(\pi^*(s)\sigma(s))^2\,ds\right. \quad (2.10)\\ &\qquad \left. +k\int_0^t \pi^*(s)\sigma(s)\,dW(s)\right)\\ &\leq K\cdot \exp\left(k\int_0^t r(s)\,ds + k\int_0^t \pi^*(s)\sigma(s)\,dW(s)\right)\\ &\leq K\cdot\exp\left(2k\int_0^t r(s)\,ds\right) + K\cdot\exp\left(2k\int_0^t \pi^*(s)\sigma(s)\,dW(s)\right). \end{aligned}$$

We now consider the integral $\int_0^t r(s)\,ds$. By the form of the short rate process, in the *Ho-Lee model* we find[11]

$$\begin{aligned} \int_0^t r(s)\,ds &= \int_0^t \left(r_0 + \int_0^s a(u)\,du + \int_0^s b\,dW(u)\right)\,ds \qquad (2.11)\\ &= r_0 t + \int_0^t\int_0^s a(u)\,du ds + b\int_0^t\int_0^s dW(u)ds\\ &= \ldots + b\int_0^t (t-u)\,dW(u). \end{aligned}$$

The dots represent a term which is deterministic and bounded on $[0, T]$. Applying Proposition 1.1, in the *Vasicek model* we obtain

$$r(t) = e^{-\alpha t}\left(r_0 + \int_0^t e^{\alpha u}\Big(\theta(u) + b\zeta(u)\Big)\,du + \int_0^t b e^{\alpha u}\,dW(u)\right).$$

Hence

[11] See Ikeda/Watanabe (1981, pp. 117ff) for the interchange of Lebesgue and Ito integrals.

$$\int_0^t r(s)\,ds = \int_0^t e^{-\alpha s}\left(r_0 + \int_0^s e^{\alpha u}\Big(\theta(u) + b\zeta(u)\Big)\,du\right)ds \qquad (2.12)$$
$$+ b\int_0^t \int_0^s e^{\alpha(u-s)}\,dW(u)ds$$
$$= \ldots + b\int_0^t \int_u^t e^{\alpha(u-s)}\,dsdW(u).$$

The dots represent a term which is deterministic and bounded on $[0, T]$.

In both cases the problem is reduced to finding an estimate for terms of the form $\exp(\int_0^t h(s)\,dW(s))$ with a deterministic and bounded function h, namely

$$\exp\left(\int_0^t h(s)\,dW(s)\right) =$$
$$\underbrace{\exp\left(\int_0^t 0.5h^2(s)\,ds\right)}_{=const.} \cdot \underbrace{\exp\left(-\int_0^t 0.5h^2(s)\,ds + \int_0^t h(s)\,dW(s)\right)}_{=:Z(t)}$$

with
$$dZ(t) = Z(t)h(t)dW(t),$$
$$Z(0) = 1.$$

Applying a result of Krylov (1980, p. 85), we find that

$$\mathrm{E}\Big(\max_{0\leq t\leq T} Z(t)\Big) < +\infty.$$

Because of (2.10) and (2.11) or (2.12), respectively, $(X^*)^k$ is bounded from above by a process of the same form as Z in both models. Therefore, property 3) is proved.

Proof of 4). Since the condition $G_{xx} < 0$ is met in both models, G is obviously a $C^{1,2}$-solution of the HJB.

Proof of 5). It is sufficient to prove that (1.29) is satisfied by all *bounded* admissible bond proportions $\pi(\cdot)$. Let $(t', x', r') \in [0, T] \times I\!R^2_+ := \{y \in I\!R^2 : y > 0\}$ and $t' \leq t \leq T$. We consider the models separately.

Ho-Lee Model

The candidate for the value function is

$$G(t, x, r) = x^\gamma \cdot \exp\left(\tfrac{1}{1-\gamma}(H_1(t) - H_1(T)) + \gamma(T - t)r\right),$$

where H_1 denotes a deterministic function which is continuously differentiable. Let K_i, $i = 1, 2, 3$, be suitable constants. As H_1, π, ζ, σ, and a are bounded functions, an application of Ito's formula yields

$$
\begin{aligned}
&G(t, X(t), r(t)) \\
&= X(t)^\gamma \cdot \exp\left(\tfrac{1}{1-\gamma}(H_1(t) - H_1(T)) + \gamma(T-t)r(t)\right) \\
&= (x')^\gamma \exp\left(\gamma \int_{t'}^{t} \pi(s)\zeta(s)\sigma(s) + r(s) - 0.5(\pi(s)\sigma(s))^2 \, ds \right. \\
&\qquad\qquad \left. + \gamma \int_{t'}^{t} \pi(s)\sigma(s) \, dW(s)\right) \\
&\quad \cdot \exp\left(\tfrac{1}{1-\gamma}(H_1(t) - H_1(T))\right) \cdot \exp\left(r(t)\gamma(T-t)\right) \\
&\le K_1 \cdot \exp\left(\gamma \int_{t'}^{t} r(s) \, ds + \gamma \int_{t'}^{t} \pi(s)\sigma(s) \, dW(s)\right) \\
&\quad \cdot \exp(\gamma T r(t)) \cdot \exp(-\gamma t r(t)) \\
&= K_1 \cdot \exp\left(\gamma \int_{t'}^{t} r(s) \, ds + \gamma \int_{t'}^{t} \pi(s)\sigma(s) \, dW(s)\right) \cdot \exp\left(\gamma T \int_{t'}^{t} dr(s)\right) \\
&\quad \cdot \exp\left(-\gamma \int_{t'}^{t} s \, dr(s) - \gamma \int_{t'}^{t} r(s) \, ds\right) \\
&= K_1 \cdot \exp\left(\gamma \int_{t'}^{t} \pi(s)\sigma(s) \, dW(s)\right) \cdot \exp\left(\gamma \int_{t'}^{t} (T-s)(a(s) \, ds + b \, dW(s))\right) \\
&\le K_2 \cdot \exp\left(\gamma \int_{t'}^{t} \pi(s)\sigma(s) + b(T-s) \, dW(s)\right) \\
&\le K_3 \cdot \exp\left(\gamma \int_{t'}^{t} \pi(s)\sigma(s) + b(T-s) dW(s) \right. \\
&\qquad\qquad \left. - 0.5\gamma^2 \int_{t'}^{t} \left(\pi(s)\sigma(s) + b(T-s)\right)^2 ds\right) \\
&=: K_3 \cdot Z(t),
\end{aligned}
$$

where Z is the unique solution of

$$
dZ(t) = Z(t)\Big(\gamma(\pi(t)\sigma(t) + b(T-t))\Big) dW(t) \quad \text{mit} \quad Z(t') = 1.
$$

Using Krylov (1980, p. 85) we arrive at

$$
\mathrm{E}\Big(\sup_{t \in [t', T]} |G(t, X(t), r(t))|^2\Big) \le K_3 \cdot \mathrm{E}\Big(\sup_{t \in [t', T]} |Z(t)|^2\Big) < \infty.
$$

We have thus proved (1.29) in the Ho-Lee model.

Vasicek Model

Our candidate for the value function is

$$
G(t, x, r) = x^\gamma \cdot \exp\left(\tfrac{1}{1-\gamma}(H_2(t) - H_2(T)) + \tfrac{\gamma}{\alpha}(1 - \exp(\alpha(t-T)))r\right),
$$

where H_2 is a continuously differentiable and deterministic function. With suitable constants K_i, $i = 1, \ldots, 6$, we find that

$$\begin{aligned}
&G(t, X(t), r(t)) \\
&\quad = X(t)^\gamma \cdot \exp\left(\tfrac{1}{1-\gamma}(H_2(t) - H_2(T)) + \tfrac{\gamma}{\alpha}(1 - \exp(\alpha(t-T)))r(t)\right) \\
&\quad \leq K_1 \cdot X(t)^\gamma \cdot \exp\left(\tfrac{\gamma}{\alpha}(1 - \exp(\alpha(t-T)))r(t)\right) \\
&\quad \leq K_2 \cdot \exp\left(\gamma \int_{t'}^{t} \pi(s)\zeta(s)\sigma(s) + r(s) - 0.5(\pi(s)\sigma(s))^2 \, ds\right. \\
&\qquad\qquad \left. + \gamma \int_{t'}^{t} \pi(s)\sigma(s) \, dW(s)\right) \cdot \exp\left(\tfrac{\gamma}{\alpha}(1 - \exp(\alpha(t-T))) \cdot r(t)\right) \\
&\quad \leq K_3 \cdot \exp\left(\gamma \int_{t'}^{t} r(s) \, ds + \gamma \int_{t'}^{t} \pi(s)\sigma(s) \, dW(s)\right) \cdot \exp\left(\tfrac{\gamma}{\alpha} r(t)\right) \\
&\qquad \cdot \exp\left(-\tfrac{\gamma}{\alpha} \exp(\alpha(t-T)) \cdot r(t)\right).
\end{aligned}$$

With the definition $f^h(t, r) := \exp(\alpha(t-T)) \cdot r$ an application of Ito's formula yields

$$\begin{aligned}
f^h(t, r(t)) &= f^h(t', r') + \int_{t'}^{t} \alpha \exp(\alpha(s-T)) r(s) \, ds + \int_{t'}^{t} \exp(\alpha(s-T)) \, dr(s) \\
&= f^h(t', r') + \int_{t'}^{t} \exp(\alpha(s-T)) \cdot (\theta(s) + b\zeta(s)) \, ds \\
&\quad + \int_{t'}^{t} b \exp(\alpha(s-T)) \, dW(s).
\end{aligned}$$

Hence, by virtue of the stochastic integral equation of the short rate, we obtain

$$\begin{aligned}
&G(t, X(t), r(t)) \\
&\quad \leq K_4 \cdot \exp\left(\gamma \int_{t'}^{t} r(s) \, ds + \gamma \int_{t'}^{t} \pi(s)\sigma(s) \, dW(s)\right) \cdot \exp\left(\tfrac{\gamma}{\alpha} r(t)\right) \\
&\qquad \cdot \exp\left(-\tfrac{\gamma}{\alpha} \int_{t'}^{t} b \exp(\alpha(s-T)) \, dW(s)\right) \\
&\quad = K_4 \cdot \exp\left(\gamma \int_{t'}^{t} r(s) \, ds + \gamma \int_{t'}^{t} \pi(s)\sigma(s) \, dW(s)\right) \\
&\qquad \cdot \exp\left(\tfrac{\gamma}{\alpha} r' + \tfrac{\gamma}{\alpha} \int_{t'}^{t} (\theta(s) - \alpha r(s) + b\zeta(s)) \, ds + \tfrac{\gamma}{\alpha} \int_{t'}^{t} b \, dW(s)\right) \\
&\qquad \cdot \exp\left(-\tfrac{\gamma}{\alpha} \int_{t'}^{t} b \exp(\alpha(s-T)) \, dW(s)\right) \\
&\quad \leq K_5 \cdot \exp\left(\int_{t'}^{t} \gamma\pi(s)\sigma(s) + \tfrac{\gamma}{\alpha} b\Big(1 - \exp(\alpha(s-T))\Big) \, dW(s)\right)
\end{aligned}$$

$$\begin{aligned}
&\leq K_6 \cdot \exp\left(\int_{t'}^{t} \gamma\pi(s)\sigma(s) + \tfrac{\gamma}{\alpha}b\Big(1-\exp(\alpha(s-T))\Big)\, dW(s)\right.\\
&\qquad\left. - \int_{t'}^{t} 0.5\left[\gamma\pi(s)\sigma(s) + \tfrac{\gamma}{\alpha}b\Big(1-\exp(\alpha(s-T))\Big)\right]^2 ds\right)\\
&=: K_6 \cdot \tilde{Z}(t).
\end{aligned}$$

Since the process $\tilde{Z}$ has the same properties as Z in the Ho-Lee model, an analogous argument leads to (1.29).

Proof of 6). By (2.9), we conclude $X^* \geq 0$.

The following proposition summarizes our results.

Proposition 2.1 (Optimal Bond Portfolio) *The optimal portfolio processes in the above bond portfolio problems are given by*

$$\pi^*(t) = \frac{1}{1-\gamma} \cdot \frac{\zeta(t)}{\sigma(t)} - \frac{\gamma}{1-\gamma} \cdot \kappa(t)$$

with

a) Ho-Lee case: $\kappa(t) = \frac{T-t}{T_1-t}$, *b) Vasicek case:* $\kappa(t) = \frac{1-e^{-\alpha(T-t)}}{1-e^{-\alpha(T_1-t)}}$.

2.2.2 Mixed Stock and Bond Portfolio Problem

In this subsection we assume that the investor can put his funds into the money market account, into stocks, and into bonds with maturity $T_1 > T$. The dynamics of these assets are given by

$$\begin{aligned}
dM(t) &= M(t)r(t)dt,\\
dS(t) &= S(t)\Big[\mu_S(t)dt + \sigma_S(t)dW_S(t) + \sigma_{SB}(t)dW_B(t)\Big],\\
dP(t) &= P(t)\Big[\underbrace{(r(t) + \zeta_B(t)\sigma_B(t))}_{=:\mu_B(t)}dt + \sigma_B(t)dW_B(t)\Big],
\end{aligned}$$

where (W_S, W_B) is a two-dimensional Brownian motion and where, for ease of notation, we write $P(t)$ instead of $P(t, T_1)$. In our model the stock price depends on two risk factors: The first factor W_S contains the specific risk of the stock, and the second W_B comes from the stochastic interest rate model.

In Merton's portfolio problem we can split the (deterministic) drift μ_S of the stock into a liquidity premium (LP) and an excess return, which should be interpreted as risk premium (RP) in this context:[12]

$$\mu_S = \underbrace{r}_{\text{LP}} + \underbrace{\mu_S - r}_{\text{RP}}.$$

The drift of the stock S under consideration can also be split

$$\mu_S(t) = r(t) + \underbrace{\mu_S(t) - r(t)}_{=:\lambda_S(t)},$$

where λ_S denotes the excess return on the stock. In the following, we assume that the excess return $\lambda_S(\cdot)$ on the stock is deterministic and continuous. This implies that $\lambda_S(\cdot)$ is progressively measurable and bounded on $[0, T]$. Furthermore, assume that the coefficients $\sigma_S(\cdot)$, $\sigma_{SB}(\cdot)$, and $\sigma_B(\cdot)$ are deterministic and continuous. In addition, let $\sigma_S(\cdot)$ and $\sigma_B(\cdot)$ be bounded away from zero.

As before we consider both the Ho-Lee and the Vasicek models:

$$dr(t) = a(t)dt + bdW_B(t)$$

with $a(t) = \tilde{a}(t) + b\zeta(t)$ in the Ho-Lee model and $a(t) = \theta(t) - \alpha r(t) + b\zeta(t)$ in the Vasicek model.

Moreover, we have $\sigma_B(t) = -b(T_1 - t)$ in the Ho-Lee model and $\sigma_B(t) = \frac{b}{\alpha}(\exp(-\alpha(T_1 - t)) - 1)$ in the Vasicek model.

In this framework the wealth equation (2.1) has the following form

$$\begin{aligned} dX(t) = X(t)\Big[&(\pi_S(t)\lambda_S(t) + \pi_B(t)\lambda_B(t) + r(t))dt \\ &+\pi_S(t)\sigma_S(t)dW_S(t) + (\pi_S(t)\sigma_{SB}(t) + \pi_B(t)\sigma_B(t))dW_B(t)\Big], \end{aligned}$$

where $\lambda_B(t) := \mu_B(t) - r(t)$ and $\pi := (\pi_S, \pi_B)$.

Using the notations of (1.1) in Chapter 1 leads to

$$\begin{aligned} Y(t) &= (X(t), r(t))', \\ \Lambda(t, x, r, \pi) &= (x(\pi_S\lambda_S + \pi_B\lambda_B + r), a)', \\ \Sigma(t, x, r, \pi) &= \begin{pmatrix} x\pi_S\sigma_S & x(\pi_S\sigma_{SB} + \pi_B\sigma_B) \\ 0 & b \end{pmatrix}, \end{aligned}$$

[12] There is no uniform use of the words excess return, risk premium and market price of risk. Apart from the above interpretation of drift, throughout this section we denote $\lambda = \mu - r$ as excess return, $\frac{\lambda}{\sigma}$ as market price of risk and $\frac{\lambda}{\sigma^2}$ as risk premium.

$$\Sigma^*(t,x,r,\pi) = \begin{pmatrix} x^2(\pi_S^2\sigma_S^2 + (\pi_S\sigma_{SB} + \pi_B\sigma_B)^2) & bx(\pi_S\sigma_{SB} + \pi_B\sigma_B) \\ bx(\pi_S\sigma_{SB} + \pi_B\sigma_B) & b^2 \end{pmatrix},$$

$$\begin{aligned} A^\pi G(t,x,r) &= G_t + 0.5x^2(\pi_S^2\sigma_S^2 + (\pi_S\sigma_{SB} + \pi_B\sigma_B)^2)G_{xx} + 0.5b^2 G_{rr} \\ &\quad + bx(\pi_S\sigma_{SB} + \pi_B\sigma_B)G_{xr} + x(\pi_S\lambda_S + \pi_B\lambda_B + r)G_x + aG_r. \end{aligned}$$

Hence, we need to solve the following HJB

$$\sup_{|\pi|\le\delta} A^\pi G(t,x,r) = 0,$$
$$G(T,x,r) = x^\gamma.$$

This will again be done by the 3-step-algorithm.

1st step: Assuming $G_{xx} < 0$ we calculate the candidates for the optimal portfolio proportions

$$\pi_S^* = -\underbrace{\left(\eta_S - \tfrac{\sigma_{SB}}{\sigma_B}\eta_{BS}\right)}_{=:\hat{\eta}_S} \cdot \frac{G_x}{xG_{xx}}, \tag{2.13}$$

$$\pi_B^* = -\underbrace{\left((1 + \tfrac{\sigma_{SB}^2}{\sigma_S^2})\eta_B - \tfrac{\sigma_{SB}}{\sigma_B}\eta_S\right)}_{=:\hat{\eta}_B} \cdot \frac{G_x}{xG_{xx}} - \frac{b}{\sigma_B} \cdot \frac{G_{xr}}{xG_{xx}} \tag{2.14}$$

with $\eta_S := \lambda_S/\sigma_S^2$, $\eta_B := \lambda_B/\sigma_B^2$ and $\eta_{BS} := \lambda_B/\sigma_S^2$.

2nd step: Substituting $\pi_S^*(t,x,r;G)$ and $\pi_B^*(t,x,r;G)$ into the HJB yields the PDE

$$\begin{aligned} 0 = G_t G_{xx} &+ \underbrace{(0.5\sigma_S^2\hat{\eta}_S^2 + 0.5(\sigma_{SB}\hat{\eta}_S + \sigma_B\hat{\eta}_B)^2 - \lambda_S\hat{\eta}_S - \lambda_B\hat{\eta}_B)}_{=:\tilde{\zeta}(t)} G_x^2 \\ &- 0.5b^2 G_{xr}^2 + 0.5b^2 G_{rr}G_{xx} - b\tfrac{\lambda_B}{\sigma_B}G_x G_{xr} + xrG_x G_{xx} + aG_r G_{xx} \end{aligned}$$

with $G(T,x,r) = x^\gamma$. This PDE is of the same form as the corresponding PDE (2.4) above.[13] Note that $\tilde{\zeta}$, in analogy to ζ in (2.4), is a continuous and deterministic function. Therefore, in the Ho-Lee model we arrive at

$$G(t,x,r) = x^\gamma \cdot \exp\left(\tfrac{1}{1-\gamma}(H_3(t) - H_3(T)) + \gamma(T-t)r\right)$$

and in the Vasicek model

$$G(t,x,r) = x^\gamma \cdot \exp\left(\tfrac{1}{1-\gamma}(H_4(t) - H_4(T)) + \tfrac{\gamma}{\alpha}(1 - \exp(\alpha(t-T)))r\right)$$

[13] PDE (2.4) follows if $\lambda_S \equiv 0$, $\sigma_S \equiv 0$, and $\sigma_{SB} \equiv 0$.

with continuously differentiable functions H_3 and H_4. Differentiating and substituting into (2.13) and (2.14) we obtain in *both* models

$$\begin{aligned}\pi_S^*(t) &= \frac{1}{1-\gamma}\cdot\left(\eta_S(t) - \tfrac{\sigma_{SB}(t)}{\sigma_B(t)}\eta_{BS}(t)\right)\\ &= \frac{1}{1-\gamma}\cdot\hat{\eta}_S(t),\\ \pi_B^*(t) &= \frac{1}{1-\gamma}\cdot\left(\left(1+\tfrac{\sigma_{SB}^2(t)}{\sigma_S^2(t)}\right)\eta_B(t) - \tfrac{\sigma_{SB}(t)}{\sigma_B(t)}\eta_S(t) - \gamma\cdot\kappa(t)\right)\\ &= \frac{1}{1-\gamma}\cdot\left(\hat{\eta}_B(t) - \gamma\cdot\kappa(t)\right),\end{aligned}$$

where $\kappa(t) = \frac{T-t}{T_1-t}$ in the *Ho-Lee model* and $\kappa(t) = \frac{1-e^{-\alpha(T-t)}}{1-e^{-\alpha(T_1-t)}}$ in the *Vasicek model.*

Both proportions are continuous and deterministic processes and thus bounded.

3rd step: By a similar line of argumentation as in Subsection 2.2.1, we can apply Corollary 1.2. Therefore in both models we must check the following assumptions

1) $\pi^*(\cdot)$ is progressively measurable,
2) $\pi^*(\cdot)$ meets condition (ii) in Definition 1.1,
3) $\pi^*(\cdot)$ meets condition (iii) in Definition 1.1,
4) G is a $C^{1,2}$-solution of the HJB,
5) condition (1.29) is met,
6) $X^{\pi^*} \geq 0$.

Note that $\pi^* := (\pi_S^*, \pi_B^*)'$.

Conditions 1) and 2) are met because in both models $\pi^*(\cdot)$ is a continuous and deterministic process. Obviously, 4) is fulfilled. Condition 6) is met because, by Proposition 1.1, we obtain

$$\begin{aligned}&X(t) =\\ &x_0 \exp\Bigg(\int_0^t \pi_S(s)\lambda_S(s) + \pi_B(s)\lambda_B(s) + r(s)\\ &\qquad - 0.5\Big((\pi_S(s)\sigma_S(s))^2 + (\pi_S(s)\sigma_{SB}(s) + \pi_B(s)\sigma_B(s))^2\Big)\,ds\\ &\qquad + \int_0^t \pi_S(s)\sigma_S(s)\,dW_S(s) + \int_0^t \pi_S(s)\sigma_{SB}(s) + \pi_B(s)\sigma_B(s)\,dW_B(s)\Bigg)\end{aligned}$$

for an admissible control $\pi(\cdot)$. Furthermore, since the wealth process has the same properties as in Subsection 2.2.1, we can prove 3) and 5) using analogous arguments.

The following proposition summarizes our results:

Proposition 2.2 (Optimal Mixed Portfolio) *The optimal portfolio processes in the above mixed portfolio problem are given by*

$$\pi_S^*(t) = \frac{1}{1-\gamma} \cdot \Big(\underbrace{\eta_S(t) - \frac{\sigma_{SB}(t)}{\sigma_B(t)}\eta_{BS}(t)}_{=:\hat{\eta}_S}\Big), \qquad \textit{(stock)}$$

$$\pi_B^*(t) = \frac{1}{1-\gamma} \cdot \Big(\underbrace{\Big(1 + \frac{\sigma_{SB}^2(t)}{\sigma_S^2(t)}\Big)\eta_B(t) - \frac{\sigma_{SB}(t)}{\sigma_B(t)}\eta_S(t)}_{=:\hat{\eta}_B} - \gamma \cdot \kappa(t)\Big) \qquad \textit{(bond)}$$

with

a) Ho-Lee case: $\kappa(t) = \frac{T-t}{T_1-t}$, *b) Vasicek case:* $\kappa(t) = \frac{1-e^{-\alpha(T-t)}}{1-e^{-\alpha(T_1-t)}}$.

Considering the optimal positions, the analogy to the pure bond problem becomes clear: The variables $\hat{\eta}_S$ as well as $\hat{\eta}_B$ can be interpreted as modified risk premiums, where both are weighted differences of η_S and η_{BS} as well as η_B and η_S. In the optimal stock position the risk premium of the stock is corrected by η_{BS}, which stands for the risk premium of the bond with respect to the stock.

Similarly, the risk premium of the bond contains a correction of the optimal bond position by the risk premium of the stock. Both these corrections are plausible ones as an increase of the risk premium of the bond makes stock investments less attractive and vice versa. Apart from that the interpretation of the bond part as given in Section 2.2.1 remains valid.

Furthermore, we obtain the optimal bond position of Subsection 2.2.1 if we choose $\sigma_S \equiv 0$ and $\sigma_{SB} \equiv 0$ in $\pi_B(\cdot)$.

However, since the detailed proof of Proposition 2.2 would have been much longer than that of Proposition 2.1, we have decided to present only the one for Proposition 2.1, as it contains the main ideas and techniques.

2.3 Dothan and Black-Karasinski Model

In this section we look at a portfolio problem in which the interest rate dynamics of the economy can be described by a Dothan or Black-Karasinski model,[14] i.e. the short rate evolves according to the SDE

$$dr(t) = r(t)\Big[a(t)dt + b(t)dW(t)\Big] \tag{2.15}$$

in the Dothan model, where we assume that a and b are measurable deterministic functions of time. Additionally, let a be integrable and let $b \neq 0$ a.s. as well as square-integrable. In the Black-Karasinski model the short rate is given by

$$d\ln(r(t)) = \kappa(t)\Big(\theta(t) - \ln(r(t))\Big)dt + \sigma(t)dW(t), \tag{2.16}$$

with measurable and deterministic functions κ, θ, $\sigma > 0$ a.s. Besides, let $\kappa \cdot \theta$ and κ be integrable and σ square-integrable. We wish to stress that both models can be embedded in the Heath-Jarrow-Morton framework. To prove this property, it is sufficient to verify Baxter's condition,[15] which reads

$$\mathrm{E}_Q\left(\int_0^t |r(s)|\, e^{-\int_0^s r(u)\,du}\, ds\right) < +\infty,$$

where Q denotes a martingale measure. Since in both models $r \geq 0$, by Fubini's theorem, we obtain

$$\mathrm{E}_Q\left(\int_0^t |r(s)|\, e^{-\int_0^s r(u)\,du}\, ds\right) \leq \mathrm{E}_Q\left(\int_0^t r(s)\, ds\right) = \int_0^t \mathrm{E}_Q\Big(r(s)\Big)\, ds < +\infty$$

because $r(t)$ is lognormally distributed. For example, if $\tilde{a}$ and b are the drift and the volatility of (2.15) under Q and both are constants, we arrive at $\mathrm{E}_Q\Big(r(s)\Big) = r_0\, e^{\tilde{a}s}$.

Suppose now that the investor maximizes utility from terminal wealth with respect to a power utility function $U(x) = \frac{1}{\gamma}x^\gamma$, $\gamma > 0$. Additionally, we assume that he can at least put funds into the money market account. One crucial problem that arises with the above interest rate models is that in contrast to affine term structure models explicit formulae of the bond prices

[14] Note that the Black-Karasinski model is the continuous limit of the Black-Derman-Toy model.

[15] See Baxter (1997).

are not available.[16] Therefore, it is not at all clear how the coefficients of the wealth equation should look like. But if the investor puts his wealth only into the money market account, clearly, his wealth equation reads as follows:

$$dX(t) = X(t)rdt, \qquad X(0) = x_0 > 0$$

and this representation is independent of all other traded assets. Without loss of generality let $x_0 = 1$ so that the wealth equation coincides with the money market account. We will now treat the models separately.

Dothan Model

The SDE (2.15) of the short rate has the following solution

$$r(t) = r_0 \exp\left(\int_0^t a(s) - 0.5b^2(s)\,ds + \int_0^t b(s)\,dW(s)\right).$$

Defining

$$m := \min_{u\in[0,t]} \exp\left(\int_0^u a(s) - 0.5b^2(s)\,ds\right)$$

and applying the Jensen inequality as well as a Fubini theorem for the interchange of Lebesgue and Ito integrals,[17] we obtain

$$\begin{aligned}
&\mathrm{E}\Big(X^\gamma(t)\Big)\\
&= \mathrm{E}\bigg(\exp\Big[\gamma \int_0^t r(s)\,ds\Big]\bigg)\\
&= \mathrm{E}\bigg(\exp\Big[\gamma r_0 \int_0^t \exp\Big\{\int_0^u a(s) - 0.5b^2(s)\,ds + \int_0^u b(s)\,dW(s)\Big\}\,du\Big]\bigg)\\
&\geq \mathrm{E}\bigg(\exp\Big[\gamma r_0 m \int_0^t \exp\Big\{\int_0^u b(s)\,dW(s)\Big\}\,du\Big]\bigg)\\
&\overset{\text{Jensen}}{\geq} \mathrm{E}\bigg(\exp\Big[\gamma r_0 m \exp\Big\{\int_0^t \int_0^u b(s)\,dW(s)du\Big\}\Big]\bigg)\\
&\overset{\text{Fubini}}{=} \mathrm{E}\bigg(\exp\Big[\gamma r_0 m \exp\Big\{\underbrace{\int_0^t \int_s^t b(s)\,dudW(s)}_{=:I}\Big\}\Big]\bigg),
\end{aligned}$$

[16] Actually, Dothan (1978) derives a semi-explicit solution of each bond price for a model with constant coefficients. Nevertheless, this formula consists of a modified Bessel function and a double integral involving hyperbolic sine and cosines so that the integral has to be solved numerically.

[17] See Ikeda/Watanabe (1981), pp. 117ff.

where

$$I \sim \mathcal{N}\Big(0, \int_0^t b^2(s)(t-s)^2 ds\Big).$$

It is a well-known result that expectations of terms like $\exp(\exp(N))$ with a normally distributed random variable N are infinite. Hence, we arrive at

$$\mathrm{E}\Big(X^\gamma(t)\Big) \geq \mathrm{E}\Big(\exp\Big(\gamma r_0 m\, e^I\Big)\Big) = +\infty.$$

We thus conclude that it is optimal for our investor to put his wealth into the money market account, i.e. "doing nothing" is optimal irrespective of whether he can additionally invest funds in stocks, bonds, or any other asset. However, recalling our findings of the portfolio problem with Vasicek term structure, this is really not a result which one would expect. Obviously, an economy where stocks and other assets are traded but where investors do not buy them cannot be in a state of equilibrium. We now proceed with the Black-Karasinski model and subsequently summarize our findings.

Black-Karasinski Model

Defining the process

$$dz(t) = \kappa(t)\Big(\theta(t) - z(t)\Big)dt + \sigma(t)dW(t),$$

the short rate is given by $r(t) = e^{z(t)}$. Then, by Proposition 1.3, we obtain $z(t) \geq y(t)$ where the dynamics of y are given by

$$dy(t) = -\kappa(t)y(t)dt + \sigma(t)dW(t).$$

Hence,

$$y(t) = \exp\Big(-\int_0^t \kappa(s)\,ds\Big)\cdot\Big[y(0) + \int_0^t \sigma(s)\exp\Big(\int_0^s \kappa(u)\,du\Big)\,dW(s)\Big].$$

Again consider the investor's expected utility

$$\mathrm{E}\Big(\exp\Big[\gamma\int_0^t r(s)\,ds\Big]\Big) \geq \mathrm{E}\Big(\exp\Big[\gamma\int_0^t e^{y(s)}\,ds\Big]\Big) \geq \mathrm{E}\Big(\exp\Big[\gamma\, e^{\int_0^t y(s)\,ds}\Big]\Big).$$

Consequently, we need to calculate the distribution of the integral $\int_0^t y(s)\,ds$:

$$\int_0^t y(s)\,ds$$

$$= \int_0^t e^{-\int_0^l \kappa(s)\,ds} \cdot \left[y(0) + \int_0^l \sigma(s) e^{\int_0^s \kappa(u)\,du}\, dW(s) \right] dl$$

$$= \underbrace{y(0) \int_0^t e^{-\int_0^l \kappa(s)\,ds} dl}_{=:\mu} + \underbrace{\int_0^t e^{-\int_0^l \kappa(s)\,ds} \int_0^l \sigma(s) e^{\int_0^s \kappa(u)\,du}\, dW(s) dl}_{=:I}.$$

Again applying the Fubini theorem for the interchange of Lebesgue and Ito integrals, we obtain

$$I = \int_0^t \int_0^l \underbrace{e^{-\int_0^l \kappa(q)\,dq} \sigma(s) e^{\int_0^s \kappa(u)\,du}}_{=:h(s,l)} dW(s) dl$$

$$= \int_0^t \int_s^t h(s,l)\, dl dW(s) \sim \mathcal{N}\Big(0, \underbrace{\int_0^t \Big[\int_s^t h(s,l)\, dl\Big]^2 ds}_{=:\nu^2}\Big),$$

where $\nu > 0$ due to our assumption $\sigma > 0$. Therefore, we end up with same result as in the Dothan model

$$\mathrm{E}\Big(\exp\Big[\gamma \int_0^t r(s)\, ds\Big]\Big) \geq \mathrm{E}\Big(\exp\Big[\gamma\, e^{\int_0^t y(s)\,ds}\Big]\Big) = \mathrm{E}\Big(\exp\Big[\gamma\, e^{\mu + I}\Big]\Big) = +\infty.$$

The following proposition summarizes our findings

Proposition 2.3 (Exploding Expectations) *If we consider a portfolio problem with the Dothan term structure (2.15) or the Black-Karasinski term structure (2.16), then we obtain for all* $\gamma > 0$

$$\mathrm{E}\Big(\exp\Big(\gamma \int_0^t r(s)\, ds\Big)\Big) = +\infty.$$

Hence, independent of other traded assets it is optimal for an investor who maximizes utility from terminal wealth with respect to a power utility function with $\gamma > 0$ *to put his wealth into the money market account.*

This proposition has important implications:

- If we assume that in the situation of Proposition 2.3 the economy can be described by a representative investor, this investor will put his wealth into the money market account. Therefore, nobody would buy bonds. Besides, although we have not introduced stocks in our portfolio problem, that can be done as in the Vasicek case. Clearly, it would not change the result of the above proposition, i.e. nobody in the economy would buy stocks either. The strange thing is that this state of affairs persists even if the economy becomes risk-seeking, i.e. $\gamma > 1$. Hence, the government can close the exchanges of the economy in the described situation.

- Although we have proved Proposition 2.3 under the physical measure, the result obviously remains valid if we assume that the dynamics of the short rates (2.15) and (2.16) are given under a risk-neutral measure. For example, the choice $\gamma = 1$ leads to an explosion of the expected value of the money market account under the risk-neutral measure. From Cox/Ingersoll/Ross (1981) we know that the prices of futures can be represented as rollover positions in the underlying. Hence, we end up with infinite prices of Eurodollar futures. This inconvenient property was first shown by Hogan/Weintraub (1993) for constant coefficients a, b or κ, θ, σ, respectively.[18] More precisely, they proved that under a risk neutral measure Q for all $0 \le u \le t \le T$

$$\mathrm{E}_Q\left(\frac{1}{P(t,T)}\bigg|\mathcal{F}_u\right) = +\infty.$$

Then, by Jensen's inequality, it follows that

$$\Big[P(t,T)\Big]^{-1} = \left[\mathrm{E}_Q\left(e^{-\int_t^T r(s)\,ds}\Big|\mathcal{F}_t\right)\right]^{-1} \le \mathrm{E}_Q\left(e^{\int_t^T r(s)\,ds}\Big|\mathcal{F}_t\right).$$

Taking conditional expectations with respect to $\mathcal{F}_u$ on both sides, we get

$$+\infty = \mathrm{E}_Q\left(\frac{1}{P(t,T)}\bigg|\mathcal{F}_u\right) \le \mathrm{E}_Q\left(e^{\int_t^T r(s)\,ds}\Big|\mathcal{F}_u\right).$$

Clearly, their proof is in no way straightforward because, for instance, they have to deal with the above-mentioned representation of the bond price in the Dothan model. For this reason we chose to prove the result using the direct way. Besides, our result is more general in the sense that we allowed for deterministic instead of constant coefficients and we did not restrict our considerations to the risk neutral measure.

Unfortunately, for investors with $\gamma < 0$ our idea as expressed above no longer works because $\mathrm{E}\Big(\exp\Big(\gamma \int_0^t r(s)\,ds\Big)\Big)$ remains finite. More precisely, this expected value is smaller than one since we are dealing with term structure

[18] See also Sandmann/Sondermann (1997).

models that provide positive short rates. Hence, it seems to be impossible to solve the problem without exactly specifying the investment opportunities. Besides, the result of Wachter (2001) that an investor will put his wealth into that bond which has a maturity equal to his investment horizon gives rise to the suggestion that the solution will probably change if $\gamma < 0$. But we have to be careful with the application of her propositions. Although Wachter (2001) proved her results under fairly general assumptions,[19] the technical requirements are unfortunately not precisely stated in her paper. In fact she assumed that the martingale method can be applied and she quoted results of Cox/Huang (1991) and Dybvig/Rogers/Back (1999) stating that the method can be applied if all moments of the state price density

$$\phi(t) := \exp\left(- \int_0^t r(s)\,ds - 0.5 \int_0^t ||\zeta(s)||^2\,ds - \int_0^t \zeta(s)\,dW(s)\right)$$

and its inverse ϕ^{-1} are finite. Here ζ denotes the market price of risk of the economy. Clearly, we cannot expect this condition to hold in the Dothan or the Black-Karasinski model. A straightforward example is an economy where $\zeta \equiv 0$. Then, from Proposition 2.3 we know that

$$\mathrm{E}\Big(\big[\phi^{-1}(t)\big]^n\Big) = \mathrm{E}\Big(\exp\Big[n\int_0^t r(s)\,ds\Big]\Big) = +\infty$$

for all $n \in I\!N$, i.e. all moments of ϕ^{-1} are infinite.

2.4 Cox-Ingersoll-Ross Model

In this section we look at the same portfolio problem as in Section 2.2 but where the interest rate model differs. We now suppose that the interest rate dynamics of the economy can be described by the Cox-Ingersoll-Ross model, i.e. the short rate evolves according to the SDE

$$dr(t) = \kappa(\theta - r(t))dt + \sigma\sqrt{r(t)}\,dW(t), \tag{2.17}$$

$t \in [0, T^*]$, κ, θ, $\sigma > 0$, with initial value $r(0) = r_0$. Besides, let

$$2\kappa\theta \geq \sigma^2 \tag{2.18}$$

[19] Wachter (2001) only assumed that the investor can buy bonds with maturities equal to the investment horizon and did not specify the utility function.

so that $P\Big(r(t) > 0, t \in [0,T]\Big) = 1$. As before, for ease of exposition, we consider a portfolio problem where the investor can allocate his wealth to a money market account M and a (zero) bond $P(\cdot, T_1)$ with maturity $T_1 > T$. In the Cox-Ingersoll-Ross model the price processes can be represented by the Ito processes

$$\begin{aligned} dM(t) &= M(t)r(t)dt, \\ dP(t,T_1) &= P(t,T_1)\Big[(r(t)(1-\lambda\bar{B}(t,T_1))dt - \bar{B}(t,T_1)\sigma\sqrt{r(t)}dW(t)\Big], \end{aligned}$$

where W is a one-dimensional Brownian motion and

$$\bar{B}(t,T_1) = \frac{2(e^{\upsilon(T_1-t)}-1)}{(\upsilon+\kappa+\lambda)(e^{\upsilon(T_1-t)}-1)+2\upsilon} > 0$$

with $\upsilon := \sqrt{(\kappa+\lambda)^2+2\sigma^2}$. The parameter λ is related to the market price of risk, $\zeta(t) = \sqrt{r(t)}\,\lambda/\sigma$. Let $\pi(t)$ be the proportion invested in the bond at time t. Then we obtain the following dynamics of the wealth equation

$$dX(t) = X(t)\Big[\Big(r(t) - \pi(t)r(t)\lambda\bar{B}(t,T_1))\Big)dt - \pi(t)\bar{B}(t,T_1)\sigma\sqrt{r(t)}dW(t)\Big] \tag{2.19}$$

with initial wealth $X(0) = x_0$. Given an investor who maximizes utility from terminal wealth at time T with respect to a power utility function $U(x) = \frac{1}{\gamma}x^\gamma$, $x \geq 0$, $\gamma \in (-\infty, 0) \cup (0,1)$, we face the following optimization problem

$$\max_{\pi(\cdot)} \mathrm{E}\Big(\tfrac{1}{\gamma}X^\pi(T)\Big)^\gamma \tag{2.20}$$

with

$$\begin{aligned} dX^\pi(t) &= X^\pi(t)\Big[\Big(r(t) - \pi(t)r(t)\lambda\bar{B}(t,T_1))\Big)dt - \pi(t)\bar{B}(t,T_1)\sigma\sqrt{r(t)}dW(t)\Big], \\ X^\pi(0) &= x_0. \end{aligned}$$

Before we tackle this problem, we first look at the portfolio problem in which the investor puts his wealth into the money market account to demonstrate that under certain circumstances the problem facing us is identical to that arising in the Dothan and the Black-Karasinski models. Suppose that the investor maximizes utility from terminal wealth with respect to a power utility function with $0 < \gamma < 1$. Then for $\pi \equiv 0$ his expected utility reads as follows

$$\mathrm{E}\bigg(\exp\Big(\gamma\int_0^t r(s)\,ds\Big)\bigg).$$

Defining $\delta := \kappa\theta$ the dynamics (2.17) of the short rate is given by

$$dr(t) = (\delta - \kappa r(t))dt + \sigma\sqrt{r(t)}\, dW(t). \tag{2.21}$$

Due to Proposition 1.3 and Remark d) of Proposition 1.2 for the auxiliary short rate

$$d\hat{r}(t) = (\tfrac{\sigma}{4} - \kappa\hat{r}(t))dt + \sigma\sqrt{r(t)}\, dW(t) \tag{2.22}$$

it follows $P\Big(\hat{r}(s) \le r(s),\, 0 \le s \le t\Big) = 1$, $t \in I\!R^+$. Note that this short rate does not meet (2.18), but this is not relevant. The reason why we introduce the process $\hat{r}$ is that, in contrast to most of the square-root processes, the SDE (2.22) has the explicit solution[20]

$$\hat{r}(t) = e^{-\kappa t}\left(\sqrt{\hat{r}(0)} + 0.5\sigma \int_0^t e^{0.5\kappa s} dW(s)\right)^2.$$

This is because $\hat{r}(t) = y^2(t)$ where y is an Ornstein-Uhlenbeck process with dynamics

$$dy(t) = -0.5\kappa y(t)dt + 0.5\sigma dW(t)$$

that due to Corollary (1.1) lead to the explicit solution

$$y(t) = e^{-0.5\kappa t}\left(y(0) + 0.5\sigma \int_0^t e^{0.5\kappa s}\, dW(s)\right).$$

Hence, applying the Hölder inequality, we arrive at

$$\begin{aligned}
&\mathrm{E}\left(\exp\left(\gamma \int_0^t r(s)\, ds\right)\right) \\
&\ge \mathrm{E}\left(\exp\left(\gamma \int_0^t \left\{e^{-0.5\kappa u}\left[y(0) + 0.5\sigma \int_0^u e^{0.5\kappa s}\, dW(s)\right]\right\}^2 du\right)\right) \\
&\ge \mathrm{E}\left(\exp\left(\gamma e^{-\kappa t} \int_0^t \left\{y(0) + 0.5\sigma \int_0^u e^{0.5\kappa s}\, dW(s)\right\}^2 du\right)\right) \\
&\overset{\text{Hölder}}{\ge} \mathrm{E}\left(\exp\left(\gamma e^{-\kappa t} t^{-1}\left[\int_0^t \underbrace{\left\{y(0) + 0.5\sigma \int_0^u e^{0.5\kappa s}\, dW(s)\right\}}_{:=\hat{X}(u)} du\right]^2\right)\right).
\end{aligned} \tag{2.23}$$

Defining

[20] From Proposition 1.2 we know that the SDE (2.21) has a unique solution, but only if the coefficients meet the condition $n := 4\delta/\sigma \in I\!N$ an explicit solution is available. This solution equals the sum of n squared and independent Ornstein-Uhlenbeck processes. See Magshoodi (1996) or Elliott/Kopp (1999), pp. 236 ff.

$$\beta(s) := \int_0^s e^{\kappa u}\, du = \kappa^{-1}(e^{\kappa s} - 1)$$

we obtain

$$\alpha(u) := \beta^{-1}(u) = \kappa^{-1} \ln(\kappa u + 1)$$

and

$$\alpha'(u) = \frac{1}{\kappa u + 1}.$$

From the theorem of time-change for martingales[21] we know that $\hat{X}(\alpha(u)) = \hat{W}(u)$, where $\hat{W}$ is a Brownian motion. Note that the Lebesgue integral in relation (2.23) is actually pathwise a Riemann integral so that we can apply the substitution rule pathwise to obtain

$$\begin{aligned}
&\mathrm{E}\Big(\exp\Big(\gamma \int_0^t r(s)\, ds\Big)\Big) \\
&\geq \mathrm{E}\Big(\exp\Big(\gamma e^{-\kappa t} t^{-1} \Big[\int_0^t \Big\{ y(0) + 0.5\sigma \hat{X}(t) \Big\}\, du \Big]^2 \Big)\Big) \\
&= \mathrm{E}\Big(\exp\Big(\gamma e^{-\kappa t} t^{-1} \Big[\int_{\beta(0)}^{\beta(t)} \Big\{ y(0) + 0.5\sigma\, \hat{X}(\alpha(u))\, \alpha'(u) \Big\}\, du \Big]^2 \Big)\Big) \\
&= \mathrm{E}\Big(\exp\Big(\gamma e^{-\kappa t} t^{-1} \Big[\int_0^{\beta(t)} \Big\{ y(0) + 0.5\sigma\, \hat{W}(u)\, \alpha'(u) \Big\}\, du \Big]^2 \Big)\Big).
\end{aligned}$$

To proceed we need the following lemma.

Lemma 2.1 *Suppose g to be a measurable deterministic function of bounded variation. We then obtain*

$$\int_0^t W(s) dg(s) = \int_0^t \Big(g(t) - g(s) \Big)\, dW(s) \sim \mathcal{N}\Big(0, \int_0^t \Big(g(t) - g(s) \Big)^2 ds\Big).$$

Consequently, for a continuous function h with primitive H we get

$$\int_0^t W(s) h(s) ds = \int_0^t \Big(H(t) - H(s) \Big)\, dW(s) \sim \mathcal{N}\Big(0, \int_0^t \Big(H(t) - H(s) \Big)^2 ds\Big).$$

Proof. To prove the first result, apply Ito's formula to $W(t) \cdot g(t)$. The second result is then obvious because

$$\int_0^t W(s) h(s)\, ds = \int_0^t W(s)\, dH(s).$$

[21] See Karatzas/Shreve (1991), p. 174.

Note that $\int_0^t W(s)dg(s)$ is pathwise (well-)defined because g has bounded variation. Besides, H is of bounded variation on $[0,t]$ because it has a continuous derivative on $[0,t]$ which is bounded on this interval. □

For fixed $t > 0$ define the random variable $Z \sim \mathcal{N}(\mu, \nu_\sigma^2)$ with $\mu = y(0)\beta(t)$ and

$$\nu_\sigma^2 = 0.25\sigma^2 \int_0^{\beta(t)} \Big(\underbrace{\alpha(\beta(t))}_{=t} - \alpha(s)\Big)^2 ds = 0.25\sigma^2 \int_0^t \Big(t-s\Big)^2 \beta'(s)\, ds$$

$$= 0.25\tfrac{\sigma^2}{\kappa^3}\Big[(1+\kappa(t-s))^2+1\Big]e^{\kappa s}\Big|_{s=0}^{s=t} = 0.25\tfrac{\sigma^2}{\kappa^3}\Big[2e^{\kappa t} - (1+\kappa t)^2 - 1\Big].$$

Hence, ν_σ is an increasing linear function of σ. Applying Lemma 2.1, we get

$$\begin{aligned}
&\mathrm{E}\Big(\exp\Big(\gamma \int_0^t r(s)\, ds\Big)\Big) \\
&\geq \mathrm{E}\Big(\exp\Big(\gamma e^{-\kappa t} t^{-1}\Big[\int_0^{\beta(t)} \Big\{y(0) + 0.5\sigma \hat{W}(u)\, \alpha'(u)\Big\}\, du\Big]^2\Big)\Big) \\
&= \mathrm{E}\Big(\exp\Big(\underbrace{\gamma e^{-\kappa t} t^{-1}}_{=:c} \cdot Z^2\Big)\Big) = \int_{-\infty}^{+\infty} \tfrac{1}{\sqrt{2\pi}\,\nu_\sigma} e^{-\frac{(z-\mu)^2}{2\nu_\sigma^2}} e^{cz^2}\, dz \\
&\geq \int_\mu^{+\infty} \tfrac{1}{\sqrt{2\pi}\,\nu_\sigma} e^{-\frac{(z-\mu)^2}{2\nu_\sigma^2}} e^{cz^2}\, dz \geq \int_\mu^{+\infty} \tfrac{1}{\sqrt{2\pi}\,\nu_\sigma} e^{-\frac{z^2}{2\nu_\sigma^2}} e^{cz^2}\, dz \\
&= \tfrac{1}{\sqrt{2\pi}\,\nu_\sigma} \int_\mu^{+\infty} e^{\frac{(2c\nu_\sigma^2 - 1)z^2}{2\nu_\sigma^2}}\, dz.
\end{aligned}$$

For fixed γ, κ, $t > 0$ we can choose $\sigma \in I\!R^+$ with[22]

$$\sigma^2 \geq \frac{2\kappa^3 t e^{\kappa t}}{\gamma[2e^{\kappa t} - (1+\kappa t)^2 - 1]} \geq \frac{\kappa^2}{2\gamma} \tag{2.24}$$

so that

$$\frac{(2c\nu_\sigma^2 - 1)}{2\nu_\sigma^2} \geq 0.$$

Then the value of the last integral becomes $+\infty$. Hence, there are parameterizations of the interest rate model that lead to

$$\mathrm{E}\Big(\tfrac{1}{\gamma}\Big(X^\pi(T)\Big)^\gamma\Big) = +\infty$$

[22] Note that $2xe^x \geq e^x - 0.5(1+x)^2 - 0.5$ for $x \geq 0$. With $x = \kappa t$ the second inequality follows.

for the riskless strategy $\pi \equiv 0$. Note that this result is valid for all $\delta > 0$. Therefore, we can choose $\delta > 0$ so that condition (2.18) is satisfied.

We emphasize that we have proved this result for all $\gamma > 0$. This means that for all $\gamma > 0$ one can find parametrizations where the above expected value is infinite for finite values of σ and δ. Consequently, "doing nothing" can once again be optimal! The following proposition summarizes our results.

Proposition 2.4 (Exploding Expectations for $\gamma > 0$) *If we consider a portfolio problem with Cox-Ingersoll-Ross term structure (2.17), then for all γ, κ, $t \in I\!R^+$, there exist σ, $\theta \in I\!R^+$ such that $2\kappa\theta \geq \sigma^2$ and*

$$\mathrm{E}\left(\exp\left(\gamma \int_0^t r(s)\, ds\right)\right) = +\infty.$$

More precisely, for γ, κ, $t \in I\!R^+$ it is sufficient to choose σ according to (2.24) and $\theta \geq \sigma^2/2\kappa$.

Obviously, for $\gamma < 0$ the above expected value is smaller than one. Hence, it is tempting to believe that in this case the portfolio problem is as well-behaved as in the Vasicek case. Unfortunately, this is wrong, as the following proposition shows. It states that for $\gamma < 0$ there exist parametrizations so that even not all bounded portfolio processes are admissible.

Proposition 2.5 (Exploding Expectations for $\gamma < 0$) *If we consider a portfolio problem with Cox-Ingersoll-Ross term structure (2.17), then for all $\gamma < 0$ and t, $\kappa \in I\!R^+$ with $\kappa > \lambda$, there exist σ, $\theta \in I\!R^+$, and a bounded portfolio process π such that $2\kappa\theta \geq \sigma^2$ and*

$$\mathrm{E}\left(\left|\tfrac{1}{\gamma}\left(X^\pi(T)\right)^\gamma\right|\right) = +\infty.$$

Proof. Fix $\gamma < 0$ and $t > 0$. The solution of the wealth equation reads as follows

$$\begin{aligned} X(t) &= x_0 \exp\left(\int_0^t r(s) - \pi(s) r(s) \lambda \bar{B}(s,T_1) - 0.5\pi^2(s)\bar{B}^2(s,T_1)\sigma^2 r(s)\, ds - \right. \\ &\quad \left. - \int_0^t \pi \bar{B}(s,T_1)\sigma\sqrt{r(s)}\, dW(s)\right) \\ &= x_0 \exp\left(\int_0^t r(s)\left[1 + 0.5\tfrac{\lambda^2}{\sigma^2} - 0.5\left(\pi(s)\bar{B}(s,T_1)\sigma + \tfrac{\lambda}{\sigma}\right)^2\right] ds \right. \\ &\quad \left. - \int_0^t \pi \bar{B}(s,T_1)\sigma\sqrt{r(s)}\, dW(s)\right). \end{aligned}$$

If we now choose $\pi(s) = c/\bar{B}(s, T_1)$ with a constant $c \in I\!R$, this leads to

$$\begin{aligned}
X^\gamma(t) &= x_0^\gamma \exp\left(\gamma \int_0^t r(s)\Big[1 - 0.5c^2\sigma^2 - c\lambda\Big]\, ds \right.\\
&\quad \left. -\gamma \int_0^t c\sigma\sqrt{r(s)}\, dW(s)\right)\\
&= x_0^\gamma \exp\left(-\gamma \int_0^t r(s)\Big[c\lambda + 0.5c^2\sigma^2 - 1\Big]\, ds \right.\\
&\quad \left. -\gamma \int_0^t c\sigma\sqrt{r(s)}\, dW(s)\right).
\end{aligned}$$

Integrating the SDE (2.17) of the short rate, we obtain

$$\int_0^t \sigma\sqrt{r(s)}\, dW(s) = r(t) - r(0) - \int_0^t \kappa(\theta - r)\, ds.$$

Hence, we arrive at

$$\begin{aligned}
X^\gamma(t) &= x_0^\gamma \exp\left(-\gamma \int_0^t r(s)\Big[c\lambda + 0.5c^2\sigma^2 - 1\Big]\, ds \right.\\
&\quad \left. -\gamma c\Big[r(t) - r(0) - \int_0^t \kappa(\theta - r)\, ds\Big]\right).\\
&\geq const \cdot \exp\left(-\gamma \int_0^t r(s)\Big[\underbrace{c(\kappa + \lambda) + 0.5c^2\sigma^2 - 1}_{=:k(c)}\Big]\, ds\right).
\end{aligned}$$

There now exists a constant $c^* > 0$ such that $k(c^*) > 0$. According to Proposition 2.4 there exists a constant σ with

$$\sigma^2 \geq \frac{2\kappa^3 t e^{\kappa t}}{|\gamma| k(c^*)[2e^{\kappa t} - (1 + \kappa t)^2 - 1]}$$

and $\theta \geq \sigma^2/2\kappa$ so that the assertion follows. □

Proposition 2.5 has the same implications as Proposition 2.3 in the Dothan or the Black-Karasinski model. But, whereas these models lead to exploding expectations of the money market account for each parametrization, the Cox-Ingersoll-Ross model has this property only for certain parameter choices. Therefore and because an explicit representation of the bond price is available, we proceed to analyze the problem using stochastic control theory.

As in the Vasicek model we need to look at a two-dimensional state process $Y = (X, r)$. Note that the second component cannot be controlled via $\pi(\cdot)$. We thus obtain

$$\begin{aligned}
Y(t) &= \Big(X(t), r(t)\Big)', \\
\Lambda(t,x,r,\pi) &= \Big(x(r - \pi r\lambda\bar{B}); \kappa(\theta - r)\Big)', \\
\Sigma(t,x,r,\pi) &= \Big(-x\pi\bar{B}\sigma\sqrt{r}; \sigma\sqrt{r}\Big)', \\
\Sigma^*(t,x,r,\pi) &= \begin{pmatrix} x^2\pi^2\bar{B}^2\sigma^2 r & -x\pi\bar{B}\sigma^2 r \\ -x\pi\bar{B}\sigma^2 r & \sigma^2 r \end{pmatrix}, \\
A^\pi G(t,x,r) &= G_t + 0.5x^2\pi^2\bar{B}^2\sigma^2 r G_{xx} - x\pi\bar{B}\sigma^2 r G_{xr} + 0.5\sigma^2 r G_{rr} \\
&\quad + x(r - \pi r\lambda\bar{B})G_x + \kappa(\theta - r)G_r.
\end{aligned}$$

Hence, the following Hamilton-Jacobi-Bellman equation (HJB) needs to be solved

$$\begin{aligned}
\sup_{|\pi|\le\delta} A^\pi G(t,x,r) &= 0, \\
G(T,x,r) &= \tfrac{1}{\gamma}x^\gamma.
\end{aligned}$$

We will see in the sequel that the optimal portfolio process is bounded. Consequently, as in the Vasicek model, we can find a $\delta > 0$ such that the absolute value of the optimal portfolio process is smaller than δ.

Due to the presence of the products rx and $\sqrt{r}x$ in the above setting, usual verification theorems, which require Lipschitz conditions, are not applicable to our problem, as both the wealth process and the short rate are unbounded processes. We therefore apply Corollary 1.3 of Chapter 1 and proceed to solve the HJB following the usual three step procedure.

We start with the calculation of the optimal bond position $\pi(\cdot)$.

1st step: Assuming $G_{xx} < 0$ we find the following candidate for the optimal bond position

$$\pi^* = \frac{\lambda}{\bar{B}\sigma^2}\frac{G_x}{xG_{xx}} + \frac{1}{\bar{B}}\frac{G_{xr}}{xG_{xx}}.$$

2nd step: Substituting $\pi^*(t,x,r;G)$ into the HJB leads to the PDE

$$\begin{aligned}
0 = {} & G_t G_{xx} + xrG_x G_{xx} - 0.5\tfrac{\lambda^2}{\sigma^2}G_x^2 r - \lambda G_{xr}G_x r \\
& + \kappa(\theta - r)G_r G_{xx} - 0.5\sigma^2 G_{xr}^2 r + 0.5\sigma^2 G_{rr}G_{xx} r
\end{aligned}$$

with terminal condition $G(T,x,r) = \frac{1}{\gamma}x^\gamma$. The form of this condition suggests the following separation ansatz

$$G(t,x,r) = \tfrac{1}{\gamma}x^\gamma \cdot \Big(f(t,r)\Big)^{1-\gamma}$$

with $f(T,r) = 1$ for all r. This leads to a second-order PDE for f of the form

$$0 = f_t + \tfrac{\gamma}{1-\gamma}\Big(1 + 0.5\tfrac{1}{1-\gamma}\tfrac{\lambda^2}{\sigma^2}\Big)rf + \Big(\kappa\theta - r\Big(\kappa - \lambda\tfrac{\gamma}{1-\gamma}\Big)\Big)f_r + 0.5\sigma^2 r f_{rr}$$

with terminal condition $f(T,r) = 1$. Assume that f is in $C^{1,2}$. Applying the Feynman-Kac representation theorem, we obtain[23]

$$f(t,r) = \tilde{\mathrm{E}}^{t,r}\Big[\exp\Big(\int_t^T \underbrace{\tfrac{\gamma}{1-\gamma}\Big(1 + 0.5\tfrac{1}{1-\gamma}\tfrac{\lambda^2}{\sigma^2}\Big)}_{=:-K} \cdot r(s)\,ds\Big)\Big]$$

with
$$dr(t) = \Big(\kappa\theta - r(t)\Big(\underbrace{\kappa - \lambda\tfrac{\gamma}{1-\gamma}}_{=:\tilde{\kappa}}\Big)\Big)dt + \sigma\sqrt{r(t)}d\tilde{W}(t) \qquad (2.25)$$
$$= \tilde{\kappa}\Big(\tfrac{\kappa\theta}{\tilde{\kappa}} - r(t)\Big)dt + \sigma\sqrt{r(t)}d\tilde{W}(t),$$

where we have changed the measure to $\tilde{P}$ which is defined by the Girsanov density

$$Z(t) := \frac{d\tilde{P}}{dP}\bigg|_{\mathcal{F}_t} = \exp\Big(-0.5(\tfrac{\gamma}{1-\gamma})^2\int_0^t \zeta^2(s)\,ds + \tfrac{\gamma}{1-\gamma}\int_0^t \zeta(s)\,dW(s)\Big)$$

with $\zeta(t) := \sqrt{r(t)}\,\lambda/\sigma$. The corresponding expectation is denoted by $\tilde{\mathrm{E}}$, and the process

$$\tilde{W}(t) = W(t) - \tfrac{\gamma}{1-\gamma}\int_0^t \zeta(s)\,ds$$

is a Brownian motion under $\tilde{P}$. Note that, by Proposition 1.2, (2.25) even has a unique solution if $\tilde{\kappa} \le 0$. The change of measure is only valid if

(i) Z is a density,

(ii) f is well-defined.

To verify these points, one needs to show that $\mathrm{E}(Z(T)) = 1$ and that the expected value in the definition of f is finite. Due to Proposition 2.4 it is not possible that both points hold in general, i.e. for all $\gamma < 1$ and all possible

[23] See Fleming/Soner (1993), pp. 400f. Note that they proved a Feynman-Kac theorem which allows $-K$ to be positive. This is contrary to the Feynman-Kac theorems which can be found in the textbooks of Karatzas/Shreve (1991, p. 268) or Korn/Korn (2001, pp. 119f).

parametrizations of the Cox-Ingersoll-Ross model. Nevertheless, we want to deduce a sufficient condition. In order to do this, let us first state and prove a variant of a theorem by Pitman/Yor (1982).

Proposition 2.6 (Pitman/Yor (1982)) *Consider the short rate (2.17) in the Cox-Ingersoll-Ross model and let*

$$\varphi(t,T,r) := \mathrm{E}\Big(\exp\Big(-\alpha r(T) - \beta \int_t^T r(s)\,ds \Big) \Big| r(t) = r \Big)$$

be the Laplace transform of $\Big(r(T), \int_t^T r(s)\,ds\Big)$. *Then* φ *is well-defined if*

$$\beta \geq -\frac{\kappa^2}{2\sigma^2} \qquad \text{and} \tag{2.26}$$

$$\alpha \geq -\frac{\kappa + a}{\sigma^2} \tag{2.27}$$

with $a := \sqrt{\kappa^2 + 2\beta\sigma^2}$. *More precisely, by defining* $k := \frac{\alpha + \frac{\kappa - a}{\sigma^2}}{\alpha + \frac{\kappa + a}{\sigma^2}}$ *we obtain*

$$\varphi(t,T,r) = \exp\Big(-A(t,T) - B(t,T)\cdot r \Big)$$

with

$$A(t,T) = \begin{cases} \frac{-\kappa\theta(\kappa-a)(T-t)}{\sigma^2} + \frac{2\kappa\theta}{\sigma^2}\ln\left(\frac{1-ke^{-a(T-t)}}{1-k}\right), & \text{if } \beta > -\frac{\kappa^2}{2\sigma^2},\ \alpha > -\frac{\kappa+a}{\sigma^2}, \\ \frac{-\kappa^2\theta(T-t)}{\sigma^2} + \frac{2\kappa\theta}{\sigma^2}\ln\left(\frac{(\alpha\sigma^2+\kappa)(T-t)}{2} + 1\right), & \text{if } \beta = -\frac{\kappa^2}{2\sigma^2},\ \alpha > -\frac{\kappa+a}{\sigma^2}, \\ \frac{-\kappa\theta(\kappa+a)(T-t)}{\sigma^2}, & \text{if } \beta \geq -\frac{\kappa^2}{2\sigma^2},\ \alpha = -\frac{\kappa+a}{\sigma^2}, \end{cases}$$

$$B(t,T) = \begin{cases} -\frac{-k(\kappa+a)e^{-a(T-t)}+\kappa-a}{\sigma^2(-ke^{-a(T-t)}+1)}, & \text{if } \beta > -\frac{\kappa^2}{2\sigma^2},\ \alpha > -\frac{\kappa+a}{\sigma^2}, \\ -\frac{\kappa}{\sigma^2} + \frac{2}{\sigma^2(T-t)+\frac{2\sigma^2}{\alpha\sigma^2+\kappa}}, & \text{if } \beta = -\frac{\kappa^2}{2\sigma^2},\ \alpha > -\frac{\kappa+a}{\sigma^2}, \\ -\frac{\kappa+a}{\sigma^2}, & \text{if } \beta \geq -\frac{\kappa^2}{2\sigma^2},\ \alpha = -\frac{\kappa+a}{\sigma^2}. \end{cases}$$

Remarks.

a) Pitman/Yor (1982) have given this proposition in a more general setting, but only for α, $\beta > 0$.

b) Deelstra/Graselli/Koehl (2000) emphasized that this domain can be extended, but they did not prove this statement. We wish to stress that there exist upper bounds for $-\alpha$ as well as for $-\beta$ and not only for $-\beta$ alone, as explicitly stated by Deelstra/Graselli/Koehl (2002) and implicitly assumed by Deelstra/Graselli/Koehl (2000). As we shall see in the following, it is far from obvious, but fortunately valid, that the bound for α is never violated in the Cox-Ingersoll-Ross portfolio problem if the bound for β is met. In general, this is wrong.

c) The values of the functions A and B in the special case $\alpha = -(\kappa + a)/\sigma^2$ are equal to the limit of A and B in the general case $\alpha > -(\kappa + a)/\sigma^2$ if k goes to infinity.

d) In the special case $\alpha = 0$ the function B reads

$$B(t,T) = \begin{cases} 2\beta \frac{e^{a(T-t)}-1}{e^{a(T-t)}(\kappa+a)-\kappa+a}, & if\ \beta > -\frac{\kappa^2}{2\sigma^2},\ 0 > -\frac{\kappa+a}{\sigma^2}, \\ -\frac{\kappa}{\sigma^2} + \frac{2}{\sigma^2(T-t)+\frac{2\sigma^2}{\kappa}}, & if\ \beta = -\frac{\kappa^2}{2\sigma^2},\ 0 > -\frac{\kappa+a}{\sigma^2}, \\ -\frac{\kappa+a}{\sigma^2}, & if\ \beta \geq -\frac{\kappa^2}{2\sigma^2},\ 0 = -\frac{\kappa+a}{\sigma^2}. \end{cases} \quad (2.28)$$

e) Note that the following proof differs from the one given by Elliot/Kopp (1999). At the beginning of their proof they have applied Ito's formula without, however, demonstrating why the corresponding function is an element of $C^{1,2}$. Moreover, they have only considered the case $\alpha > 0$ and $\beta > 0$.

f) We wish to stress that for the following proof it is irrelevant whether κ is positive or not. Therefore, Proposition 2.6 is valid for all $\kappa \in I\!R$.

Proof.[24] We start with the ansatz

$$\varphi(t,T,r) = \exp\Big(- A(t,T) - B(t,T) \cdot r \Big),$$

where $A(\cdot,T)$ and $B(\cdot,T)$ are twice continuous differentiable functions of time t, and show that

$$e^{Y(t)} := e^{-\beta \int_0^t r(s)\,ds - A(t,T) - B(t,T) r(t)}$$

is a P-martingale. Fixing $0 \leq u \leq t$, we obtain

$$\begin{aligned} \mathrm{E}\Big(e^{Y(t)}\Big|\mathcal{F}_u\Big) &= \mathrm{E}\Big(e^{-\beta \int_0^t r(s)\,ds - A(t,T) - B(t,T) r(t)}\Big|\mathcal{F}_u\Big) \\ &= \mathrm{E}\Big(e^{-\beta \int_0^t r(s)\,ds}\varphi(t,T,r(t))\Big|\mathcal{F}_u\Big) \end{aligned}$$

[24] My thanks to Chris Rogers for pointing out the idea of the proof in the special case when α is zero.

$$
\begin{aligned}
&= \mathrm{E}\left(e^{-\beta\int_0^t r(s)\,ds}\mathrm{E}\left(e^{-\alpha r(T)-\beta\int_t^T r(s)\,ds}\,\middle|\, r(t)\right)\middle|\mathcal{F}_u\right)\\
&\overset{\text{Markov}}{=} \mathrm{E}\left(e^{-\beta\int_0^t r(s)\,ds}\mathrm{E}\left(e^{-\alpha r(T)-\beta\int_t^T r(s)\,ds}\,\middle|\, \mathcal{F}_t\right)\middle|\mathcal{F}_u\right)\\
&= e^{-\beta\int_0^u r(s)\,ds}\mathrm{E}\left(\mathrm{E}\left(e^{-\alpha r(T)-\beta\int_u^T r(s)\,ds}\,\middle|\, \mathcal{F}_t\right)\middle|\mathcal{F}_u\right)\\
&= e^{-\beta\int_0^u r(s)\,ds}\mathrm{E}\left(e^{-\alpha r(T)-\beta\int_u^T r(s)\,ds}\,\middle|\,\mathcal{F}_u\right)\\
&\overset{\text{Markov}}{=} e^{-\beta\int_0^u r(s)\,ds}\mathrm{E}\left(e^{-\alpha r(T)-\beta\int_u^T r(s)\,ds}\,\middle|\,r(u)\right)\\
&= e^{-\beta\int_0^u r(s)\,ds}\varphi(u,T,r(u))\\
&= e^{-\beta\int_0^u r(s)\,ds}e^{-A(u,T)-B(u,T)r(u)} = e^{Y(u)}.
\end{aligned}
$$

The equations marked "Markov" are valid due to the Markov property of the process $\{r(t)\}_t$. Note that all above conditional expectations are well-defined because their arguments are positive. Applying Ito's formula to the process Y, we obtain

$$
\begin{aligned}
&de^{Y(t)}\\
&= e^{Y(t)}dY(t) + 0.5e^{Y(t)}d < Y >_t\\
&= e^{Y(t)}\Big[-\beta r(t)dt - A_t(t,T)dt - B_t(t,T)r(t)dt\\
&\qquad -B(t,T)\Big(\kappa(\theta - r(t))dt + \sigma\sqrt{r(t)}dW(t)\Big)\Big] + 0.5e^{Y(t)}B^2(t,T)\sigma^2 r(t)dt\\
&= e^{Y(t)}\Big[\Big\{ -\beta - B_t(t,T) + B(t,T)\kappa + 0.5B^2(t,T)\sigma^2\Big\}r(t)\\
&\qquad +\Big\{ -A_t(t,T) - B(t,T)\kappa\theta\Big\}\Big]dt - e^{Y(t)}B(t,T)\sigma\sqrt{r(t)}dW(t).
\end{aligned}
$$

Hence, e^Y is only a martingale if the drift of the above SDE vanishes. This leads to the following differential equations

$$-\beta - B_t(t,T) + B(t,T)\kappa + 0.5B^2(t,T)\sigma^2 = 0, \tag{2.29}$$

$$-A_t(t,T) - \kappa\theta B(t,T) = 0 \tag{2.30}$$

for A and B with $A(T,T) = 0$ and $B(T,T) = \alpha$. Note that the differential equation for B is a Riccati equation. In order to solve it, we make the ansatz

$$B(t,T) = c\,\frac{\phi_t(t,T)}{\phi(t,T)},$$

where the constant c is to be chosen later. Computing the derivative and substituting it into (2.29) leads to

$$-\beta + c\tfrac{\phi_t}{\phi}\kappa - c\tfrac{\phi_{tt}}{\phi} + c(0.5c\sigma^2 + 1)\tfrac{\phi_t^2}{\phi} = 0$$

Choosing $c = -2/\sigma^2$ we obtain

$$\phi_{tt} - \kappa\phi_t + \tfrac{\beta}{c}\phi = 0. \tag{2.31}$$

Solving the corresponding polynomial equation

$$z^2 - \kappa z + \tfrac{\beta}{c} = 0,$$

we obtain

$$z_{1/2} = 0.5\kappa \pm 0.5\underbrace{\sqrt{\kappa^2 + 2\beta\sigma^2}}_{=:a}.$$

These solutions are only real numbers if $a \geq 0$. This leads to condition (2.26). First consider the case $a > 0$. Then every solution of (2.31) can be written as

$$\phi(t) = w_1 e^{0.5(\kappa+a)t} + w_2 e^{0.5(\kappa-a)t}$$

with w_1, $w_2 \in I\!R$. Now w_1 and w_2 need to be determined according to the terminal condition $B(T,T) = \alpha$. Due to our ansatz $B(t,T) = c\phi_t(t,T)/\phi(t,T)$ we get

$$B(t,T) = -\frac{w_1(\kappa+a)e^{0.5(\kappa+a)t} + w_2(\kappa-a)e^{0.5(\kappa-a)t}}{\sigma^2(w_1 e^{0.5(\kappa+a)t} + w_2 e^{0.5(\kappa-a)t})}. \tag{2.32}$$

Hence, we need to solve

$$\alpha = -\frac{w_1(\kappa+a)e^{0.5(\kappa+a)T} + w_2(\kappa-a)e^{0.5(\kappa-a)T}}{\sigma^2(w_1 e^{0.5(\kappa+a)T} + w_2 e^{0.5(\kappa-a)T})}$$

or, equivalently,

$$w_1\left(\alpha + \tfrac{\kappa+a}{\sigma^2}\right)e^{0.5aT} = -w_2\left(\alpha + \tfrac{\kappa-a}{\sigma^2}\right)e^{-0.5aT} \tag{2.33}$$

for w_1 or w_2. If $\alpha = -(\kappa+a)/\sigma^2$, we have $w_2 = 0$ and, consequently, $B(t,T) = -(\kappa+a)/\sigma^2$ as well as $A(t,T) = -\kappa\theta\frac{\kappa+a}{\sigma^2}(T-t)$.
Assuming $\alpha \neq -(\kappa+a)/\sigma^2$, we obtain

$$w_1 = -w_2\underbrace{\frac{\alpha + \frac{\kappa-a}{\sigma^2}}{\alpha + \frac{\kappa+a}{\sigma^2}}}_{=:k} e^{-aT}.$$

Plugging this result into (2.32), it follows that

$$B(t,T) = -\frac{-k(\kappa + a)e^{-a(T-t)} + \kappa - a}{\sigma^2(-ke^{-a(T-t)} + 1)}.$$

If $k \geq 1$, the denominator has a null and, consequently, B is not well-defined. We therefore need conditions which exclude $k \geq 1$. Since $a > 0$, we have

$$\alpha + \tfrac{\kappa - a}{\sigma^2} < \alpha + \tfrac{\kappa + a}{\sigma^2}.$$

If $\alpha + \frac{\kappa+a}{\sigma^2} > 0$, we get

$$k = \frac{\alpha + \frac{\kappa - a}{\sigma^2}}{\alpha + \frac{\kappa + a}{\sigma^2}} < 1,$$

but if $\alpha + \frac{\kappa+a}{\sigma^2} < 0$, we obtain $k > 1$. Hence, we have to require (2.27) that leads to a well-defined function B. Integrating the differential equation (2.30), we arrive at

$$A(t,T) = -\tfrac{\kappa\theta(\kappa - a)}{\sigma^2}(T - t) + \tfrac{2\kappa\theta}{\sigma^2}\ln\left(\tfrac{1 - ke^{-a(T-t)}}{1-k}\right),$$

which is well-defined due to (2.27).

Finally, let us discuss the case $a = 0$, which is equivalent to $\beta = -0.5\kappa^2/\sigma^2$. Then all solutions of (2.31) can be written as

$$\phi(t) = w_1 e^{0.5\kappa t} + w_2 t e^{0.5\kappa t}$$

with $w_1, w_2 \in I\!R$. This leads to

$$B(t,T) = -\frac{w_1\kappa + 2w_2 + w_2 t\kappa}{\sigma^2(w_1 + w_2 t)}. \tag{2.34}$$

Using the terminal condition $B(T,T) = \alpha$, we obtain

$$w_1(\alpha\sigma^2 + \kappa) = -w_2(\alpha T\sigma^2 + 2 + T\kappa).$$

If $\alpha = -\kappa/\sigma^2$, we obtain the same result as in the case when $a > 0$ and $\alpha = -(\kappa + a)/\sigma^2$. Therefore, we assume that $\alpha > -\kappa/\sigma^2$. It follows that

$$w_1 = -\left(T + \tfrac{2}{\alpha\sigma^2 + \kappa}\right)w_2.$$

Plugging this into (2.34) leads to

$$B(t,T) = -\frac{\kappa}{\sigma^2} + \frac{2}{\sigma^2(T-t) + \frac{2\sigma^2}{\alpha\sigma^2 + \kappa}}.$$

Integrating (2.30) we obtain

$$A(t,T) = -\tfrac{\kappa^2\theta}{\sigma^2}(T-t) + \tfrac{2\kappa\theta}{\sigma^2}\ln\Big(0.5(\alpha\sigma^2+\kappa)(T-t)+1\Big).$$

This completes the proof. □

We now come back to the points (i) and (ii) on page 52. Let us first discuss (i). Due to the dynamics of the short rate (2.17) we get

$$\int_0^t \sigma\sqrt{r(s)}\,dW(s) = r(t) - r(0) - \int_0^t \kappa(\theta - r(s))\,ds. \tag{2.35}$$

Hence, we obtain $Z(t) = R(t)\cdot\exp\Big(-\tfrac{\gamma}{1-\gamma}\tfrac{\lambda}{\sigma^2}(r(0)+\kappa\theta t)\Big)$, where

$$R(t) := \exp\left(\tfrac{\gamma}{1-\gamma}\tfrac{\lambda}{\sigma^2}r(t) + \int_0^t \left[\tfrac{\gamma}{1-\gamma}\tfrac{\lambda}{\sigma^2}\kappa - 0.5\Big(\tfrac{\gamma}{1-\gamma}\tfrac{\lambda}{\sigma}\Big)^2\right] r(s)\,ds\right).$$

Hence, Z is a density if

$$\mathrm{E}(R(T)) = \exp\Big(\tfrac{\gamma}{1-\gamma}\tfrac{\lambda}{\sigma^2}(r(0)+\kappa\theta T)\Big) \tag{2.36}$$

because then $\mathrm{E}(Z(T)) = 1$. We can apply Proposition 2.6 to R if

$$\beta_1 \geq -\tfrac{\kappa^2}{2\sigma^2}, \tag{2.37}$$

$$-\tfrac{\gamma}{1-\gamma}\tfrac{\lambda}{\sigma^2} \geq -\tfrac{\kappa+a_1}{\sigma^2} \tag{2.38}$$

with $\beta_1 = -\left[\tfrac{\gamma}{1-\gamma}\tfrac{\lambda}{\sigma^2}\kappa - 0.5\Big(\tfrac{\gamma}{1-\gamma}\tfrac{\lambda}{\sigma}\Big)^2\right]$ and $a_1 = \sqrt{\kappa^2 + 2\beta_1\sigma^2}$. Since

$$\begin{aligned}\tfrac{\kappa^2}{2\sigma^2} - \left[\tfrac{\gamma}{1-\gamma}\tfrac{\lambda}{\sigma^2}\kappa - 0.5\Big(\tfrac{\gamma}{1-\gamma}\tfrac{\lambda}{\sigma}\Big)^2\right] &= \tfrac{1}{2\sigma^2}\Big(\kappa^2 - 2\tfrac{\gamma}{1-\gamma}\lambda\kappa + (\tfrac{\gamma}{1-\gamma})^2\lambda^2\Big)\\ &= \tfrac{1}{2\sigma^2}\Big(\kappa - \tfrac{\gamma}{1-\gamma}\lambda\Big)^2,\end{aligned}$$

the inequality (2.37) is met. The second inequality (2.38) can be written as $a_1 \geq \tfrac{\gamma}{1-\gamma}\lambda - \kappa$. Since $a_1 \geq 0$, it is obviously valid for $\tfrac{\gamma}{1-\gamma}\lambda - \kappa \leq 0$. Besides, $a_1^2 = (\tfrac{\gamma}{1-\gamma}\lambda - \kappa)^2$. Hence, (2.38) is satisfied. Consequently, Proposition 2.6 can be applied to R and it can easily be verified that (2.36) is valid. Hence, Z is a density in any case.[25]

Let us note that it is tempting to consider Novikov's condition when analyzing whether Z is a density or not. The Novikov condition reads as follows:

$$\mathrm{E}\left(0.5(\tfrac{\gamma}{1-\gamma})^2\int_0^t \tfrac{\lambda^2}{\sigma^2}\,r(s)\,ds\right) < \infty.$$

[25] An alternative proof of this result can be found in Shirakawa (2002).

Due to Proposition 2.6, it is satisfied if

$$-0.5(\tfrac{\gamma}{1-\gamma})^2\tfrac{\lambda^2}{\sigma^2} \geq -\tfrac{\kappa^2}{2\sigma^2}.$$

For $\gamma > 0$ this inequality is equivalent to $\kappa \geq \gamma(\kappa + |\lambda|)$. Obviously, the last inequality is not met for some cases of $\gamma > 0$. Hence, the Novikov condition is too weak for our considerations.

Finally, we look at (ii). We wish to underline the fact that we are now working under the measure $\tilde{P}$. In order to apply Proposition 2.6, we need to check if (2.26) and (2.27) are met, i.e. in this context

$$K \geq -\frac{\tilde{\kappa}^2}{2\sigma^2}, \tag{2.39}$$

$$0 \geq -\frac{\tilde{\kappa} + \tilde{a}}{\sigma^2}, \tag{2.40}$$

where $\tilde{a} = \sqrt{\tilde{\kappa}^2 + 2K\sigma^2}$. The first inequality (2.39) is valid if $\gamma < 0$ or if $\gamma > 0$ and $\kappa^2 - \gamma(\kappa + \lambda)^2 - 2\sigma^2\gamma \geq 0$. This inequality is equivalent to

$$\frac{\gamma}{1-\gamma}\left(1 + \frac{\kappa\lambda}{\sigma^2} + \frac{\lambda^2}{2\sigma^2}\right) \leq \frac{\kappa^2}{2\sigma^2}. \tag{2.41}$$

Condition (2.40) is obviously met if $\gamma > 0$ and $\lambda \leq 0$ or if $\gamma < 0$ and $\lambda \geq 0$. Given $\gamma > 0$ and $\lambda > 0$, inequality (2.40) follows from (2.39) due to (2.41). If $\gamma < 0$ and $\lambda < 0$, one has to distinguish two cases: For $\frac{\gamma}{1-\gamma}\lambda - \kappa \leq 0$ condition (2.40) is clearly met. If $\frac{\gamma}{1-\gamma}\lambda - \kappa > 0$, inequality (2.40) is equivalent to $2K\sigma^2 \geq 0$, which is valid because $K > 0$ for $\gamma < 0$. Hence, (ii) is fulfilled if (2.41) is required.

In the following we assume that for $\gamma > 0$ condition (2.41) holds as strict inequality, i.e.

$$\frac{\gamma}{1-\gamma}\left(1 + \frac{\kappa\lambda}{\sigma^2} + \frac{\lambda^2}{2\sigma^2}\right) < \frac{\kappa^2}{2\sigma^2}. \tag{2.42}$$

Then (i) and (ii) are met. By Proposition 2.6, the function f is given by

$$f(t,r) = \exp\Big(-A(t,T) - B(t,T)r\Big),$$

where A and B are deterministic functions. Notice that f is indeed an element of $C^{1,2}$. So we arrive at

$$\begin{aligned} G(t,x,r) &= \tfrac{1}{\gamma}x^\gamma(f(t,r))^{1-\gamma} \\ &= \tfrac{1}{\gamma}x^\gamma \exp\Big(-(1-\gamma)\cdot A(t,T) - (1-\gamma)\cdot B(t,T)\cdot r\Big) \end{aligned}$$

as candidate for the value function and

$$\pi^*(t) = -\frac{1}{1-\gamma}\frac{\lambda}{\sigma^2 \bar{B}(t,T_1)} + \frac{B(t,T)}{\bar{B}(t,T_1)}$$

as corresponding candidate for the optimal bond position.

To solve our portfolio problem, we now apply Corollary 1.3. Going through its assumptions, we realize that every requirement is obviously met except for property (1.25) which is satisfied if for some $\rho > 1$

$$\sup_{p\in\mathbb{N}} \mathrm{E}\Big(|G(\theta_p, X^*(\theta_p), r(\theta_p))|^\rho\Big) < \infty, \tag{2.43}$$

where X^* denotes the wealth process corresponding to π^* and $\{\theta_p\}_{p\in\mathbb{N}}$ denotes some sequence of stopping times with $0 \le \theta_p \le T$. Recall that

$$\begin{aligned} dX(t) &= X(t)\Big[\Big(r(t) - \pi(t)r(t)\lambda\bar{B}(t,T_1))\Big)dt - \pi(t)\bar{B}(t,T_1)\sigma\sqrt{r(t)}dW(t)\Big], \\ dr(t) &= \kappa(\theta - r(t))dt + \sigma\sqrt{r(t)}\, dW(t). \end{aligned}$$

Hence,

$$\begin{aligned} X(t) &= x_0 \exp\Bigg(\int_0^t r(s)\Big\{1 + 0.5\tfrac{\lambda^2}{\sigma^2} - 0.5\Big[\pi(s)\bar{B}(s,T_1)\sigma + \tfrac{\lambda}{\sigma}\Big]^2\Big\}\, ds \\ &\qquad - \int_0^t \pi(s)\bar{B}(s,T_1)\sigma\sqrt{r(s)}\, dW(s)\Bigg), \\ r(t) &= r_0 + \int_0^t \kappa(\theta - r(s))\, ds + \int_0^t \sigma\sqrt{r(s)}\, dW(s). \end{aligned}$$

Besides, we obtain $\pi^*(t)\bar{B}(t,T_1)\sigma = -\frac{1}{1-\gamma}\frac{\lambda}{\sigma} + B(t,T)\sigma$. Applying Ito's formula we get

$$\begin{aligned} B(t,T)r(t) &= B(0,T)r(0) + \int_0^t B(s,T)\, dr(s) + \int_0^t r(s)\, dB(s,T) \\ &= B(0,T)r(0) + \int_0^t \Big[B(s,T)\kappa(\theta - r(s)) + r(s)B_t(s,T)\Big]\, ds \\ &\quad + \int_0^t B(s,T)\sigma\sqrt{r(s)}\, dW(s). \end{aligned}$$

Then, by (2.35), we arrive at

$$\begin{aligned} X^*(t) = x_0 \exp\Bigg(&\int_0^t r(s)\Big[1 + 0.5\tfrac{\lambda^2}{\sigma^2} - 0.5\Big(B(s,T)\sigma - \tfrac{\gamma}{1-\gamma}\tfrac{\lambda}{\sigma}\Big)^2\Big]\, ds \\ &+ \tfrac{1}{1-\gamma}\tfrac{\lambda}{\sigma^2}\Big[r(t) - r(0) - \int_0^t \kappa(\theta - r(s))\, ds\Big] + B(0,T)r(0) \\ &- B(t,T)r(t) + \int_0^t \Big[r(s)B_t(s,T) + B(s,T)\kappa(\theta - r(s))\Big]\, ds\Bigg). \end{aligned}$$

On the basis of this result we obtain

$$\begin{aligned}
&G(t, X^*(t), r(t)) \\
&= \tfrac{1}{\gamma}\Big(X^*(t)\Big)^{\gamma} \exp\Big(- (1-\gamma)A(t,T) - (1-\gamma)B(t,T)r(t)\Big) \\
&= \tfrac{1}{\gamma} x_0^{\gamma} \exp\Big(\gamma \int_0^t r(s)\Big[1 + 0.5\tfrac{\lambda^2}{\sigma^2} - 0.5\Big(B(s,T)\sigma - \tfrac{\gamma}{1-\gamma}\tfrac{\lambda}{\sigma}\Big)^2\Big]\,ds \\
&\qquad + \tfrac{\gamma}{1-\gamma}\tfrac{\lambda}{\sigma^2}\Big[r(t) - r(0) - \int_0^t \kappa(\theta - r(s))\,ds\Big] + \gamma B(0,T)r(0) \\
&\qquad - \gamma B(t,T)r(t) + \gamma \int_0^t \Big[r(s)B_t(s,T) + B(s,T)\kappa(\theta - r(s))\Big]\,ds\Big) \\
&\quad \cdot \exp\Big(- (1-\gamma)A(t,T) - (1-\gamma)B(t,T)r(t)\Big) \\
&= det(t) \cdot \exp\Big(\Big[\tfrac{\gamma}{1-\gamma}\tfrac{\lambda}{\sigma^2} - B(t,T)\Big]r(t) \\
&\qquad + \int_0^t \gamma\Big[1 + 0.5\tfrac{\lambda^2}{\sigma^2} - 0.5\Big(B(s,T)\sigma - \tfrac{\gamma}{1-\gamma}\tfrac{\lambda}{\sigma}\Big)^2 \\
&\qquad\qquad + B_t(s,T) - B(s,T)\kappa + \tfrac{1}{1-\gamma}\tfrac{\lambda}{\sigma^2}\kappa\Big]\, r(s)\,ds\Big),
\end{aligned}$$

where "$det(t)$" stands for a deterministic term which is bounded on $[0,T]$ and therefore irrelevant to our further considerations. Note that for $\gamma > 0$ it is positive. To prove (2.43), we need results which are summarized in the following lemma.

Lemma 2.2 *The following equality holds*

$$\begin{aligned}
&\gamma\Big[1 + 0.5\tfrac{\lambda^2}{\sigma^2} - 0.5\Big(B(s,T)\sigma - \tfrac{\gamma}{1-\gamma}\tfrac{\lambda}{\sigma}\Big)^2 + B_t(s,T) - B(s,T)\kappa + \tfrac{1}{1-\gamma}\tfrac{\lambda}{\sigma^2}\kappa\Big] \\
&\quad = \tfrac{\gamma}{1-\gamma}\Big(1 + \tfrac{\kappa\lambda}{\sigma^2} + \tfrac{\lambda^2}{2\sigma^2}\Big). \qquad (2.44)
\end{aligned}$$

Besides, we have

$$\tfrac{\gamma}{1-\gamma}\tfrac{\lambda}{\sigma^2} - B(t,T) \le \tfrac{\gamma}{1-\gamma}\tfrac{\lambda}{\sigma^2} \quad \text{for } \gamma < 0, \qquad (2.45)$$

$$\tfrac{\gamma}{1-\gamma}\tfrac{\lambda}{\sigma^2} - B(t,T) < \tfrac{\kappa}{\sigma^2} \quad \text{for } \gamma > 0, \qquad (2.46)$$

where for $\gamma > 0$ it is assumed that (2.41) holds as strict inequality.

Proof. First we prove (2.44):

$$\begin{aligned}
&\gamma\Big[1 + 0.5\tfrac{\lambda^2}{\sigma^2} - 0.5\Big(B(s,T)\sigma - \tfrac{\gamma}{1-\gamma}\tfrac{\lambda}{\sigma}\Big)^2 + B_t(s,T) - B(s,T)\kappa + \tfrac{1}{1-\gamma}\tfrac{\lambda}{\sigma^2}\kappa\Big] \\
&\quad = \gamma\Big[1 + 0.5\tfrac{\lambda^2}{\sigma^2} - 0.5\tfrac{\gamma^2}{(1-\gamma)^2}\tfrac{\lambda^2}{\sigma^2} + \tfrac{1}{1-\gamma}\tfrac{\lambda}{\sigma^2}\kappa - K
\end{aligned}$$

$$+K - 0.5B^2(s,T)\sigma^2 + B_t(s,T) - B(s,T)\underbrace{(\kappa - \lambda\tfrac{\gamma}{1-\gamma})}_{=\tilde{\kappa}}\Big]$$

$$\underbrace{\qquad\qquad\qquad\qquad\qquad\qquad\qquad\qquad}_{\overset{(2.29)}{=}0}$$

$$= \gamma\Big[1 + 0.5\tfrac{\lambda^2}{\sigma^2} - 0.5\tfrac{\gamma^2}{(1-\gamma)^2}\tfrac{\lambda^2}{\sigma^2} + \tfrac{1}{1-\gamma}\tfrac{\lambda}{\sigma^2}\kappa + \tfrac{\gamma}{1-\gamma}\Big(1 + 0.5\tfrac{1}{1-\gamma}\tfrac{\lambda^2}{\sigma^2}\Big)\Big]$$

$$= \tfrac{\gamma}{1-\gamma}\Big(1 + \tfrac{\kappa\lambda}{\sigma^2} + \tfrac{\lambda^2}{2\sigma^2}\Big).$$

Note that B meets (2.29) with $\beta = K$ and $\kappa = \tilde{\kappa}$.

We now prove (2.45) and (2.46). Since we require that (2.41) holds as strict inequality, the inequalities (2.39) and (2.40) are also strict. For $\gamma < 0$ this is valid in all cases. If $\gamma > 0$, condition (2.41) is equivalent to (2.39) and therefore (2.39) holds as strict inequality. Besides, recalling our arguments on page 59, we see that inequality (2.40) is at least strict if (2.39) is strict. Therefore, only the first representation of B in (2.28) is relevant to our further considerations, i.e.

$$B(t,T) = 2K\frac{e^{\tilde{a}(T-t)} - 1}{e^{\tilde{a}(T-t)}(\tilde{\kappa} + \tilde{a}) - \tilde{\kappa} + \tilde{a}}$$

with $\tilde{a} = \sqrt{\tilde{\kappa}^2 + 2K\sigma^2}$ and $K = -\tfrac{\gamma}{1-\gamma}\Big(1 + 0.5\tfrac{1}{1-\gamma}\tfrac{\lambda^2}{\sigma^2}\Big)$. Thus, if we assume $\gamma < 0$, by (2.40), we obtain $-B \leq 0$, which proves (2.45).

Now assume that $\gamma > 0$. First note that $\tilde{\kappa} \geq 0$ because (2.40) holds and $K < 0$. Besides, according to (2.40), we get $-B \geq 0$. The function B can be written as

$$B(t,T) = 2K\frac{1}{\tilde{\kappa} + \tilde{a}q(t)},$$

where q is a deterministic function with $q(t) \geq 1$. Moreover, $K > -\frac{\tilde{\kappa}^2}{2\sigma^2}$. Therefore, we obtain

$$-B(t,T) = 2(-K)\frac{1}{\tilde{\kappa} + \tilde{a}q(t)} \leq 2(-K)\frac{1}{\tilde{\kappa} + \tilde{a}} < \frac{\tilde{\kappa}}{\sigma^2}$$

because $\tilde{\kappa} \geq 0$. It follows

$$\tfrac{\gamma}{1-\gamma}\tfrac{\lambda}{\sigma^2} - B(t,T) < \tfrac{\gamma}{1-\gamma}\tfrac{\lambda}{\sigma^2} + \tfrac{\tilde{\kappa}}{\sigma^2} = \tfrac{1}{\sigma^2}\big(\tfrac{\gamma}{1-\gamma}\lambda + \kappa - \lambda\tfrac{\gamma}{1-\gamma}\big) = \tfrac{\kappa}{\sigma^2},$$

which proves (2.46). □

Define $\varepsilon = \rho - 1 > 0$, i.e. ε can be some arbitrary small positive number. First assume that $\gamma > 0$. Then, by (2.44) and (2.46), we get

$$
\begin{aligned}
&|G(t, X^*(t), r(t))|^{\rho} \\
&\overset{(2.44)}{=} det^{\rho}(t) \cdot \exp\Bigg(\rho\Big[\tfrac{\gamma}{1-\gamma}\tfrac{\lambda}{\sigma^2} - B(t,T)\Big] r(t) \\
&\qquad\qquad + \rho \int_0^t \tfrac{\gamma}{1-\gamma}\Big[1 + \tfrac{\kappa\lambda}{\sigma^2} + \tfrac{\lambda^2}{2\sigma^2}\Big] r(s)\, ds\Bigg) \qquad (2.47) \\
&\overset{(2.46)}{\leq} det^{\rho}(t) \cdot \exp\Bigg(\rho\tfrac{\kappa}{\sigma^2} r(t) + \rho \int_0^t \tfrac{\gamma}{1-\gamma}\Big[1 + \tfrac{\kappa\lambda}{\sigma^2} + \tfrac{\lambda^2}{2\sigma^2}\Big] r(s)\, ds\Bigg) \\
&= det^{\rho}(t) \cdot \exp\Bigg(\rho\tfrac{\kappa}{\sigma^2}\Big[r_0 + \int_0^t \kappa(\theta - r(s))\, ds + \int_0^t \sigma\sqrt{r(s)}\, dW(s)\Big] \\
&\quad + \rho \int_0^t \tfrac{\gamma}{1-\gamma}\Big[1 + \tfrac{\kappa\lambda}{\sigma^2} + \tfrac{\lambda^2}{2\sigma^2}\Big] r(s)\, ds\Bigg) \\
&= det_2(t) \cdot \exp\Bigg(-\rho\tfrac{\kappa^2}{\sigma^2} \int_0^t \kappa r(s)\, ds + \rho\tfrac{\kappa}{\sigma} \int_0^t \sqrt{r(s)}\, dW(s) \\
&\quad + \rho \int_0^t \tfrac{\gamma}{1-\gamma}\Big[1 + \tfrac{\kappa\lambda}{\sigma^2} + \tfrac{\lambda^2}{2\sigma^2}\Big] r(s)\, ds\Bigg) \\
&= det_2(t) \cdot \exp\Bigg(\rho\tfrac{\kappa}{\sigma} \int_0^t \sqrt{r(s)}\, dW(s) \\
&\quad + \rho \int_0^t \Big\{ \underbrace{\tfrac{\gamma}{1-\gamma}\Big[1 + \tfrac{\kappa\lambda}{\sigma^2} + \tfrac{\lambda^2}{2\sigma^2}\Big] - 0.5\tfrac{\kappa^2}{\sigma^2}(1-\varepsilon)}_{\overset{(2.42)}{\leq} 0} - 0.5\tfrac{\kappa^2}{\sigma^2}\underbrace{(1+\varepsilon)}_{=\rho} \Big\} r(s)\, ds\Bigg) \\
&\leq det_2(t) \cdot \exp\Bigg(\rho\tfrac{\kappa}{\sigma} \int_0^t \sqrt{r(s)}\, dW(s) - 0.5\rho^2\tfrac{\kappa^2}{\sigma^2} \int_0^t r(s)\, ds\Bigg), \qquad (2.48)
\end{aligned}
$$

where "$det_2(t)$" stands for a positive deterministic term which is bounded on $[0,T]$. The stochastic part of (2.48) is (at least) a local martingale and, as it is positive, a supermartingale. By the optional sampling theorem (OS),[26] we obtain for all stopping times $\theta_p \in [0,T]$

$$
\begin{aligned}
&\mathrm{E}\Big(|G(\theta_p, X^*(\theta_p), r(\theta_p))|^{\rho}\Big) \\
&\leq \mathrm{E}\Bigg(det_2(\theta_p) \cdot \exp\Big(\rho\tfrac{\kappa}{\sigma} \int_0^{\theta_p} \sqrt{r(s)}\, dW(s) - 0.5\rho^2\tfrac{\kappa^2}{\sigma^2} \int_0^{\theta_p} r(s)\, ds\Big)\Bigg) \\
&\leq \sup_{t\in[0,T]} det_2(t) \cdot \mathrm{E}\Bigg(\exp\Big(\rho\tfrac{\kappa}{\sigma} \int_0^{\theta_p} \sqrt{r(s)}\, dW(s) - 0.5\rho^2\tfrac{\kappa^2}{\sigma^2} \int_0^{\theta_p} r(s)\, ds\Big)\Bigg) \\
&\overset{OS}{\leq} \sup_{t\in[0,T]} det_2(t) < \infty
\end{aligned}
$$

[26] See e.g. Karatzas/Shreve (1991), p. 19.

and this proves (2.43) for $\gamma > 0$. Now assume that $\gamma < 0$. Then, by (2.44) and (2.45), we get

$$
\begin{aligned}
&|G(t, X^*(t), r(t))|^\rho \\
&\overset{(2.44)}{=} |det(t)|^\rho \cdot \exp\Bigg(\rho\Big[\tfrac{\gamma}{1-\gamma}\tfrac{\lambda}{\sigma^2} - B(t,T)\Big]r(t) \\
&\qquad\qquad + \rho\int_0^t \tfrac{\gamma}{1-\gamma}\Big[1 + \tfrac{\kappa\lambda}{\sigma^2} + \tfrac{\lambda^2}{2\sigma^2}\Big] r(s)\,ds\Bigg) \\
&\overset{(2.45)}{\leq} |det(t)|^\rho \cdot \exp\Bigg(\rho\tfrac{\gamma}{1-\gamma}\tfrac{\lambda}{\sigma^2} r(t) + \rho\int_0^t \tfrac{\gamma}{1-\gamma}\Big[1 + \tfrac{\kappa\lambda}{\sigma^2} + \tfrac{\lambda^2}{2\sigma^2}\Big] r(s)\,ds\Bigg) \\
&= |det(t)|^\rho \cdot \exp\Bigg(\rho\tfrac{\gamma}{1-\gamma}\tfrac{\lambda}{\sigma^2}\Big[r_0 + \int_0^t \kappa(\theta - r(s))\,ds + \int_0^t \sigma\sqrt{r(s)}\,dW(s)\Big] \\
&\quad + \rho\int_0^t \tfrac{\gamma}{1-\gamma}\Big[1 + \tfrac{\kappa\lambda}{\sigma^2} + \tfrac{\lambda^2}{2\sigma^2}\Big] r(s)\,ds\Bigg) \\
&\leq det_2(t) \cdot \exp\Bigg(- \rho\tfrac{\gamma}{1-\gamma}\tfrac{\lambda}{\sigma^2}\kappa \int_0^t r(s)\,ds + \rho\tfrac{\gamma}{1-\gamma}\tfrac{\lambda}{\sigma}\int_0^t \sqrt{r(s)}\,dW(s) \\
&\quad + \rho\int_0^t \tfrac{\gamma}{1-\gamma}\Big[1 + \tfrac{\kappa\lambda}{\sigma^2} + \tfrac{\lambda^2}{2\sigma^2}\Big] r(s)\,ds\Bigg) \\
&= det_2(t) \cdot \exp\Bigg(\rho\tfrac{\gamma}{1-\gamma}\tfrac{\lambda}{\sigma}\int_0^t \sqrt{r(s)}\,dW(s) + \rho\int_0^t \tfrac{\gamma}{1-\gamma}\Big[1 + \tfrac{\lambda^2}{2\sigma^2}\Big] r(s)\,ds\Bigg) \\
&= det_2(t) \cdot \exp\Bigg(- 0.5\rho^2(\tfrac{\gamma}{1-\gamma})^2\tfrac{\lambda^2}{\sigma^2}\int_0^t r(s)\,ds + \rho\tfrac{\gamma}{1-\gamma}\tfrac{\lambda}{\sigma}\int_0^t \sqrt{r(s)}\,dW(s) \\
&\quad + \rho\int_0^t 0.5\rho(\tfrac{\gamma}{1-\gamma})^2\tfrac{\lambda^2}{\sigma^2} + \tfrac{\gamma}{1-\gamma}\Big[1 + \tfrac{\lambda^2}{2\sigma^2}\Big] r(s)\,ds\Bigg).
\end{aligned}
$$

Since $\frac{\gamma}{1-\gamma} \in (-1, 0)$, there exists $\rho > 1$ such that $\rho\frac{\gamma}{1-\gamma} \in (-1, 0)$. Hence,

$$0.5\rho(\tfrac{\gamma}{1-\gamma})^2\tfrac{\lambda^2}{\sigma^2} + \tfrac{\gamma}{1-\gamma}\Big[1 + \tfrac{\lambda^2}{2\sigma^2}\Big] \leq 0.5\tfrac{\gamma}{1-\gamma}\tfrac{\lambda^2}{\sigma^2}\Big(\rho\tfrac{\gamma}{1-\gamma} + 1\Big) \leq 0.$$

To this end we obtain

$$
\begin{aligned}
|G(\theta_p, X^*(\theta_p), r(\theta_p))|^\rho \leq \sup_{t\in[0,T]} det_2(t) \cdot \exp\Bigg(&- 0.5\rho^2(\tfrac{\gamma}{1-\gamma})^2\tfrac{\lambda^2}{\sigma^2}\int_0^{\theta_p} r(s)\,ds \\
&+ \rho\tfrac{\gamma}{1-\gamma}\tfrac{\lambda}{\sigma}\int_0^{\theta_p} \sqrt{r(s)}\,dW(s)\Bigg).
\end{aligned}
$$

Again, by applying the optional sampling theorem, we get the desired result. Therefore, we arrive at the following proposition:

Proposition 2.7 (Optimal Bond Portfolio) *Suppose that the interest rate dynamics of an economy can be described by the Cox-Ingersoll-Ross model*

(2.17) and that an investor maximizes utility from terminal wealth at time T with respect to a power utility function $U(x) = \frac{1}{\gamma}x^{\gamma}$. If condition (2.42) is satisfied, then the proportion

$$\pi^*(t) = \frac{1}{1-\gamma}\frac{-\lambda}{\sigma^2\bar{B}(t,T_1)} + \frac{B(t,T)}{\bar{B}(t,T_1)},$$

invested in a bond with maturity $T_1 > T$ is optimal among all weakly admissible portfolio strategies. Here

$$\bar{B}(t,T_1) = \frac{2(e^{\upsilon(T_1-t)}-1)}{(\upsilon+\kappa+\lambda)(e^{\upsilon(T_1-t)}-1)+2\upsilon},$$

$$B(t,T) = 2K\frac{e^{\tilde{a}(T-t)}-1}{e^{\tilde{a}(T-t)}(\tilde{\kappa}+\tilde{a})-\tilde{\kappa}+\tilde{a}}$$

with $\upsilon = \sqrt{(\kappa+\lambda)^2+2\sigma^2}$, $K = -\frac{\gamma}{1-\gamma}\left(1+0.5\frac{1}{1-\gamma}\frac{\lambda^2}{\sigma^2}\right)$, and $\tilde{a} = \sqrt{\tilde{\kappa}^2+2K\sigma^2}$.

The form of the optimal bond position is very similar to the Ho-Lee or Vasicek model. The main difference results from condition (2.42). Clearly, from a practical point of view it is rather awkward that this condition has to hold because the investor cannot influence whether (2.42) is violated or not. We thus draw the following conclusion: Given the condition holds there exists an optimal bond portfolio, otherwise even the passive strategy to put all funds into the money market account can be optimal. Obviously, this is a negative feature of the Cox-Ingersoll-Ross model and it is at least debatable whether their model is suitable for handling interest rate risk in portfolio problems.

2.5 Widening the Investment Universe

In this section an investment problem will be considered in which the investor can in addition put funds into foreign stocks and foreign bonds. To describe this problem formally, we use the model of Amin/Jarrow (1991). The aim of this section is not to solve this extended problem but to show that such a problem can be handled as if it were a purely domestic investment problem.

For simplicity, the domestic and the foreign short rate r and r_f are modeled by a one-factor model

$$dr(t) = a(t)dt + bdW_1(t), \qquad r(0) = r_0,$$

$$dr_f(t) = a_f(t)dt + b_f dW_3(t), \quad r_f(0) = r_{0,f},$$

where $b, b_f > 0$ and the drift $a(t) = \tilde{a}(t) + b\zeta_B(t)$ or $a(t) = \theta(t) - \alpha r(t) + b\zeta(t)$ leads to the models by Ho/Lee (1986) or Vasicek (1977). The variables W_1 and W_3 denote components of a four-dimensional Brownian motion (W_1, W_2, W_3, W_4).

As before, the domestic investment opportunities - for simplicity, a bond, a stock, and a money market account - possess the following dynamics[27]

$$\begin{aligned}
dP(t,T_1) &= P(t,T_1)\Big[(r(t) + \zeta_1(t)\sigma_B(t))dt + \sigma_B(t)\,dW_1(t)\Big],\\
P(0,T_1) &= p_0(T_1),\\
dS(t) &= S(t)\Big[(r(t) + \lambda_S(t))dt + \sigma_{SB}dW_1(t) + \sigma_{SS}dW_2(t)\Big],\\
S(0) &= s_0,\\
dM(t) &= r(t)M(t)dt,\\
M(0) &= 1.
\end{aligned}$$

Note that W_1 is a risk factor which influences the domestic bond market. Hence, the stock price depends on this factor and on W_2 which models the specific risk of the stock.

On the foreign capital market the investor can trade in a bond, a stock, and a money market account which follow the dynamics

$$\begin{aligned}
dP_f(t,T_2) &= P_f(t,T_2)\Big[(r_f(t) + \zeta_3(t)\sigma_{B,f}(t))dt + \sigma_{B,f}(t)\,dW_3(t)\Big],\\
P_f(0,T_2) &= p_{0,f}(T_2),\\
dS_f(t) &= S_f(t)\Big[(r_f(t) + \lambda_{S,f}(t))dt + \sigma_{SB,f}dW_3(t) + \sigma_{SS,f}dW_4(t)\Big],\\
S_f(0) &= s_{0,f},\\
dM_f(t) &= r_f(t)M_f(t)dt\\
M_f(0) &= 1.
\end{aligned}$$

Note that $\zeta = (\zeta_1, \zeta_2, \zeta_3, \zeta_4)$ corresponds to the market price of risk with respect to the four risk factors. Therefore

$$\begin{aligned}
\lambda_S &= \zeta_1\sigma_{SB} + \zeta_2\sigma_{SS}\\
\lambda_{S_f} &= \zeta_3\sigma_{SB,f} + \zeta_4\sigma_{SS,f}
\end{aligned}$$

The exchange rate dynamics are governed by

[27] See Section 2.2.2.

$$dF(t) = F(t)\Big[\Big(r(t) - r_f(t) + \sum_{i=1}^{4} \zeta_i(t)\sigma_{i,f}(t)\Big)dt + \sum_{i=1}^{4} \sigma_{i,f}(t)dW_i(t)\Big],$$

$F(0) = f_0$, and it is quoted in units of the domestic currency (per unit of the foreign currency). Assume that $\sigma_{i,f}$, $i = 1, \ldots, 4$, are constants. Since we wish to consider a domestic investor, the exchange rate is needed to compute the domestic prices $S_f^* := S_f \cdot F$, $P_f^* := P_f \cdot F$, and $M_f^* := M_f \cdot F$ of the foreign assets. Ito's formula leads to

$$\begin{aligned}
dP_f^*(t,T_2) &= P_f^*(t,T_2)\Big[(r(t) + \lambda_{B,f}^*(t))dt + \sum_{i=1}^{4} \sigma_{B,i,f}^*(t)dW_i(t)\Big],\\
P_f^*(0,T_2) &= p_{0,f}(T_2) \cdot f_0,\\
dS_f^*(t) &= S_f^*(t)\Big[(r(t) + \lambda_{S,f}^*(t))dt + \sum_{i=1}^{4} \sigma_{S,i,f}^*(t)dW_i(t)\Big],\\
S_f^*(0) &= s_{0,f} \cdot f_0,\\
dM_f^*(t) &= M_f^*(t)\Big[(r(t) + \sum_{i=1}^{4} \zeta_i(t)\sigma_{i,f}(t))dt + \sum_{i=1}^{4} \sigma_{i,f}(t)dW_i(t)\Big],\\
M_f^*(0) &= f_0,
\end{aligned}$$

with

$$\begin{aligned}
\lambda_{B,f}^* &= \sum_{i=1}^{4} \zeta_i\sigma_{i,f} + \zeta_3\sigma_{B,f} + \sigma_{B,f}\sigma_{3,f},\\
\sigma_{B,i,f}^* &= \sigma_{i,f}, \quad i \neq 3,\\
\sigma_{B,3,f}^* &= \sigma_{3,f} + \sigma_{B,f},\\
\lambda_{S,f}^* &= \lambda_{S,f} + \sigma_{3,f}\sigma_{SB,f} + \sigma_{4,f}\sigma_{SS,f}\\
\sigma_{S,i,f}^* &= \sigma_{i,f}, \quad i = 1, 2,\\
\sigma_{S,3,f}^* &= \sigma_{3,f} + \sigma_{SB,f},\\
\sigma_{S,4,f}^* &= \sigma_{4,f} + \sigma_{SS,f}.
\end{aligned}$$

From this it is obvious that foreign assets can be treated as domestic assets because their SDEs have the same form as the SDEs of domestic assets. Therefore, all our earlier results still apply if we assume that the market prices of risk are deterministic. Clearly, the portfolio problems are more difficult to solve as usually Brownian motions of higher dimensions are involved. The following proposition summarizes our findings.

Proposition 2.8 (International Portfolio Problem)
Given a portfolio problem with domestic and foreign assets as described above,

then the foreign assets can be treated as if they were domestic assets. Therefore, the problem can be solved as if it were a purely domestic investment problem.

2.6 Conclusion

In this section we have discussed Merton's portfolio problem under the assumptions that the investor maximizes utility from terminal wealth with respect to a power utility function and that interest rates are stochastic. More precisely, we have assumed that the interest rate dynamics can be described by the models of Ho/Lee (1986), Vasicek (1977), Dothan (1978), Black/Karasinski (1991), or Cox/Ingersoll/Ross (1985). Thus, we have analyzed the classical, most relevant one-factor short rate models which are known from literature. The results are remarkable: Only in the Gaussian short rate models of Ho and Lee or Vasicek do we find an optimal portfolio process which is valid for all parametrizations of the interest rate model and all risk preferences of the investor. The lognormal models of Dothan as well as Black and Karasinski are the extreme opposites because in these models the optimal investor's utility is infinite for every parametrization of the model. This result gets worse, since it is an "optimal" strategy to put all funds into the money market account. Whereas in the theory of option pricing the drawbacks of lognormal models are well-known from Hogan/Weintraub (1993) and Sandmann/Sondermann (1997), our findings underline that these models are also useless when it comes to considering portfolio problems with stochastic interest rates. Therefore, our results on portfolio optimization are the analogs to their results on contingent claim pricing.

Further, we have shown that it is difficult to deal with the model of Cox, Ingersoll, and Ross because there are parametrizations of their interest rate model which lead to exploding expectations of the investor's utility. Nevertheless, for most parametrizations of the model a similar result as in the Gaussian models holds. From a theoretical point of view, this dichotomy should occur because the short rate of the Cox-Ingersoll-Ross model is, roughly speaking, the square of a Gaussian random variable. But for practical applications such a property seems to be disadvantageous.

Besides, our results show that some authors, e.g. Lioui/Poncet (2001), are incorrect when they apply the martingale method to portfolio problems with stochastic interest rates without checking the assumptions of this method. In general, as for example in the interest rate models with exploding utility, these assumptions are not met and, therefore, it is incorrect to believe that for stochastic interest rates the dynamic and the static optimization problems, as defined by Cox/Huang (1989, 1991), are equivalent in any case.

We have concluded this section by analyzing portfolio problems with foreign assets. Fortunately, these problems can be treated as if they were purely domestic portfolio problems.

3

Elasticity Approach to Portfolio Optimization

3.1 Introduction

Portfolio optimization has been one of the most heavily researched areas in finance dating back to the work by Markowitz (1952) who used a discrete one period model to develop his theory. Merton (1969, 1971) was the first to solve portfolio problems in a continuous-time setting by applying stochastic control techniques. Using ideas presented in the seminal papers by Harrison/Kreps (1979) and Harrison/Pliska (1981, 1983), an alternative approach - the so-called martingale approach - was developed by Pliska (1986), Cox/Huang (1989, 1991), and Karatzas/Lehoczky/Shreve (1987). In each of these papers the portfolio problems were formulated with respect to the investment opportunities of an investor. In this chapter it will be shown that an investor actually optimizes over elasticities and - in the case of stochastic interest rates - over durations. Using this *elasticity approach to portfolio optimization* (EAPO), one can apply a kind of *two-step procedure* to solve portfolio problems. Firstly, one determines the optimal elasticity, which in a complete market is independent of a specific asset. Secondly, a strategy is computed which tracks this elasticity. Our method proves to be especially useful when portfolio problems with contingent claims are studied. It allows to focus on so-called *reduced portfolio problems.* We wish to emphasize that our approach is applicable both to the stochastic control approach and the martingale approach. Nevertheless, using a two-step procedure can be interpreted as a parallel to the martingale approach.

The remainder of this chapter is structured as follows. In Section 3.2 we introduce the elasticity approach assuming that interest rates are deterministic. In Section 3.3 this method is generalized to stochastic interest rates. Section 3.4 concludes.

3.2 Elasticity in Portfolio Optimization

The control variables of Merton's portfolio problem are the proportions of the investor's wealth invested in his investment opportunities. The aim of this section is to show that a continuous-time portfolio problem can be solved as if it were an optimization problem over elasticities. In general, elasticity measures the relative response of a claim price or the portfolio value, respectively, to a small percentage change in a risk factor affecting the portfolio.

Let us consider a frictionless financial market which consists of a continuously traded stock and a money market account. The prices of these financial instruments evolve according to the stochastic differential equations

$$\begin{aligned} dS(t) &= S(t)\Big[\mu dt + \sigma dW(t)\Big], \qquad && S(0) = s_0, \\ dM(t) &= rM(t)dt, && M(0) = 1. \end{aligned}$$

Here, the short rate r, the drift μ, and the volatility σ of the stock are constants and W denotes a one-dimensional Brownian motion defined on a filtered probability space $(\Omega, \mathcal{F}, P)$. The filtration $\{\mathcal{F}_t\}_{t\geq 0}$ is the usual P-augmentation of the natural filtration of W. Assume $\mu > r$ so that an investor gains an incentive for buying stocks.

In addition, an option $C(t) = C(t, S(t))$ on the stock is being traded. An application of Ito's formula yields the dynamics

$$\begin{aligned} dC &= C_t dt + C_s dS + 0.5 C_{ss} d < S > \\ &= (C_t + C_s S\mu + 0.5 C_{ss} S^2\sigma^2)dt + C_s S\sigma dW \end{aligned} \tag{3.1}$$

of the claim, where the subscripts denote the partial derivatives with respect to time t and price s of the underlying. Note that for ease of exposition we sometimes omit the functional dependencies on time t. The well-known partial differential equation (PDE) of Black/Scholes (1973) reads as follows:

$$C_t + s(\mu - \zeta\sigma)C_s + 0.5 s^2\sigma^2 C_{ss} - rC = 0,$$

where ζ stands for the market price of risk of the stock market. Since we assume that the stock is being traded, we conclude that $\zeta = (\mu - r)/\sigma$. But dropping this assumption does not invalidate our results. Let $\lambda := \mu - r$ be the excess return on the stock. Plugging this into (3.1) results in

$$\begin{aligned} dC &= (rC + C_s S\lambda)dt + C_s S\sigma dW \\ &= C\Big[(r + \varepsilon_C \lambda)dt + \varepsilon_C \sigma dW\Big], \end{aligned}$$

where

$$\varepsilon_C = \frac{dC/C}{dS/S} := \frac{C_s S}{C}$$

denotes the elasticity of the call with respect to the stock price. Let $\varphi_S(t)$, $\varphi_C(t)$, and $\varphi_M(t)$ be the number of shares, calls, and portions of the money market account, respectively, which an investors holds at time t. Hence, his wealth is given by

$$X(t) := \varphi_S(t)S(t) + \varphi_C(t)C(t) + \varphi_M(t)M(t).$$

The wealth process of a self-financing portfolio strategy $(\varphi_S, \varphi_C, \varphi_M)$ follows the stochastic differential equation

$$dX(t) = \varphi_S(t)dS(t) + \varphi_C(t)dC(t) + \varphi_M(t)dM(t). \tag{3.2}$$

Alternatively, we can model the investor's strategy via the portfolio process $\pi = (\pi_S, \pi_C)$ which is defined by

$$\pi_S := \frac{\varphi_S S}{X}, \qquad \pi_C := \frac{\varphi_C C}{X}.$$

The fractions π_S and π_C denote the percentage of total wealth invested in stocks and calls. The percentage invested in the money market account is given by the residuum $1 - \pi_S - \pi_C$. Using these definitions we can rewrite the wealth equation (3.2) as follows:

$$\begin{aligned} dX &= \varphi_S dS + \varphi_C dC + \varphi_M dM \\ &= \varphi_S S\Big[\mu dt + \sigma dW\Big] + \varphi_C C\Big[(r + \varepsilon_C\lambda)dt + \varepsilon_C \sigma dW\Big] + \varphi_M M r dt \\ &= \pi_S X\Big[\mu dt + \sigma dW\Big] + \pi_C X\Big[(r + \varepsilon_C\lambda)dt + \varepsilon_C \sigma dW\Big] \\ &\quad + (1 - \pi_S - \pi_C)X r dt \\ &= X\Big[(r + (\pi_S + \pi_C\varepsilon_C)\lambda)dt + (\pi_S + \pi_C\varepsilon_C)\sigma dW\Big]. \end{aligned}$$

Obviously, the last equation involves the elasticity of the call with respect to the stock price. Note that for the corresponding elasticity of the stock and the money market account we have $\varepsilon_S \equiv 1$ and $\varepsilon_M \equiv 0$, respectively. Therefore, the term

$$\varepsilon := \pi_S \varepsilon_S + \pi_C \varepsilon_C = \pi_S + \pi_C \varepsilon_C$$

coincides with the static elasticity of the investor's portfolio. The word "static" emphasizes that ε only equals the elasticity of the portfolio if π is held constant. Otherwise, Ito's rule has to be applied and additional terms come into play. However, for our purposes the static elasticity is the relevant term to consider. Hence, we can rewrite the wealth equation to obtain

$$dX = X\Big[(r + \varepsilon\lambda)dt + \varepsilon\sigma dW\Big]. \tag{3.3}$$

In (3.3) we shall interpret the static portfolio elasticity ε as the control variable of the portfolio problem. Since the wealth equation of a portfolio problem in which the investment opportunities consist of a stock and a money market account reads

$$dX = X\Big[(r + \pi_S\lambda)dt + \pi_S\sigma dW\Big],$$

the optimal portfolio elasticity can be concluded from results by Merton (1969, 1971).[1] For example, if the investor maximizes utility from terminal wealth at time T with respect to the power utility function $U(x) = \frac{1}{\gamma}x^\gamma$, $x \geq 0$, $\gamma < 1$, the optimal elasticity is given by

$$\varepsilon^*(t) = \frac{1}{1-\gamma} \cdot \frac{\lambda}{\sigma^2} = const,$$

where the limiting case $\gamma = 0$ implies a logarithmic utility function. As a result, the investor has the opportunity to choose any combination of stock, call, and money market account which guarantees a portfolio elasticity of ε^*. It is important to emphasize that even in those cases when the portfolio process π stochastically varies over time the investor optimizes his portfolio over its *static* elasticity. This is possible because in a continuous-time framework he can trade at any time.

If the investor does not invest his funds in the stock, the optimal proportion invested in the call is uniquely determined and reads as follows:

[1] See also Fleming/Rishel (1975), pp. 160f, Duffie (1992), pp. 145ff, Fleming/Soner (1993), pp. 174ff or Korn (1997), pp. 48ff.

$$\pi_C^*(t) = \frac{1}{1-\gamma} \cdot \frac{\zeta}{\sigma} \cdot \frac{1}{\varepsilon_C(t)}.$$

This optimal proportion is in line with results of Korn/Trautmann (1999), who analyzed portfolio problems with path-independent options and a deterministic term structure using the martingale method.

Note that the optimal portfolio elasticity does not depend on a specific asset. Consequently, we could enlarge the set of investment opportunities by any number of contingent claims on the stock without affecting our results. However, it must be ensured that at any point in time up to the planning horizon T the investor can put funds into the money market account and into a contingent claim which is sensitive to the stock price. Hence, provided that there are tradable contingent claims, an investor can split up his portfolio decision process in two steps. Firstly, he determines his optimal portfolio elasticity. Secondly, he chooses a portfolio whose elasticity matches his optimal elasticity. So the whole problem amounts to computing elasticity. This can be done as if the investment opportunities consisted only of a stock and a money market account. Hence, every investment problem with contingent claims can be reduced to this simplified problem which we call a *reduced portfolio problem.*

So far, we have only considered an investor who maximizes utility from terminal wealth. Nevertheless, our findings remain valid if the investor chooses additionally a consumption rate $c(t)$, because then the wealth equation (3.3) reads as follows:

$$dX = X\Big[(r + \varepsilon\lambda)dt + \varepsilon\sigma dW\Big] - cdt.$$

The optimal portfolio elasticity now equals the optimal proportion invested in the stock given that the investor has the opportunity to consume or to invest in the stock or the money market account. Additionally, we conclude that the optimal consumption pattern remains the same as in Merton's portfolio problem. So each result of Merton (1969, 1971) is still valid in our particular framework. Besides, a portfolio problem with an infinite horizon can be solved similarly by choosing the portfolio elasticity and the consumption rate as control variables. Again one may adopt Merton's results. The following proposition summarizes our findings.

Proposition 3.1 (Two-Step Portfolio Selection)
Given a financial market which consists of a money market account and contingent claims on a stock. We assume that at any point in time up to the

planning horizon the investor can put his wealth into the money market account and at least one contingent claim which is sensitive to the stock price. Then the following results apply:

(i) The optimal wealth process in Merton's portfolio problem is uniquely determined by the optimal portfolio elasticity.

(ii) The optimal elasticity can be calculated without any further knowledge of the investment opportunities and equals the optimal proportion of stocks in the reduced portfolio problem.

(iii) If the investor maximizes utility from terminal wealth and/or consumption, the optimal consumption patterns remain the same as in the portfolio problems of Merton (1969, 1971).

Remarks.

a) We wish to emphasize that the stock itself can be interpreted as a contingent claim on the stock.

b) Note that the assumptions of Proposition 3.1 ensure that the market is complete according to Harrison/Kreps (1979) and Harrison/Pliska (1981, 1983). In general, a market is complete if every contingent claim is attainable. Hence, in our formulation of the problem the market is complete if the set of admissible elasticities is not restricted. It should be stressed that even in incomplete markets - for example markets with short-selling restrictions - it is still possible to apply the elasticity approach to portfolio optimization. But then the investor is confronted with constraints on the attainable elasticities. Besides, the optimal elasticity depends on the traded assets.

Clearly, the above results generalize to multidimensional Brownian motions. The following two examples highlight our findings.

Example 3.1: "Hedging vs. Portfolio Optimization"

Consider an investor who maximizes utility from terminal wealth with respect to the power utility function $U(x) = \frac{1}{\gamma}x^\gamma$. Given this investor's desire to shorten a European call on one stock, i.e. $\varphi_C^* \equiv -1$, and to put his remaining wealth into stocks, his optimal portfolio elasticity equals

$$\varepsilon^*(t) = \frac{1}{1-\gamma} \cdot \frac{\lambda}{\sigma^2}.$$

Hence, the optimal fraction π_S^* invested in the stock is given by the equation

$$\frac{1}{1-\gamma} \cdot \frac{\lambda}{\sigma^2} = \pi_S^*(t) + \varepsilon_C(t)\pi_C^*(t).$$

Since $\pi_S^* = \varphi_S^* \cdot S/X$ and $\pi_C^* = -C/X$ we obtain

$$\varphi_S^*(t) = \underbrace{C_s(t)}_{\text{"hedge"}} + \underbrace{\frac{1}{1-\gamma} \cdot \frac{\lambda}{\sigma^2} \cdot \frac{X(t)}{S(t)}}_{\text{"speculation"}}.$$

This result reveals the difference between hedging and portfolio optimization: If the investor only wished to hedge the call, he would buy $C_s(t)$ stocks which corresponds to the dynamic hedging strategy of Black/Scholes (1973). In the context of portfolio optimization this strategy is only optimal for an investor with extreme risk aversion, that means for $\gamma = -\infty$. In any other case the speculation term does not vanish, meaning that he takes risk according to his risk aversion. Note that the speculation term is always positive if λ and σ are positive. It increases with the investor's wealth because a power utility function implies decreasing absolute risk aversion (DARA).

Example 3.2: "Portfolio Optimization with Lookback Options"
Consider an investor who wishes to invest his funds in the money market account and in a European lookback call with payoff

$$C_{min}(T_{min}) = \max\{S(T_{min}) - M(T_{min}); 0\} = S(T_{min}) - M(T_{min})$$

at time $T_{min} = 1$, where $M(t) := \min_{0 \le s \le t} S(s)$ denotes the running minimum of $S(t)$ and the process $S(t)$ follows the SDE

$$dS(t) = S(t)\Big[\mu dt + \sigma dW(t)\Big]$$

with $S(0) = 100$. From Goldman/Sosin/Gatto (1979) it is well-known that in the special case $r = 0.5\sigma^2$ the price of the lookback call at time t equals

$$C_{min}(t) = P(t; M(t)) + C(t; M(t)),$$

where $P(t; M(t))$ $(C(t; M(t)))$ denotes the value of an ordinary European put (call) at time t with exercise price $M(t)$ and the same maturity as the lookback call. For ease of exposition, we choose $r = 0.02$ and $\sigma = 0.2$ so that this result can be applied. Further, let $\mu = 0.06$. In Figure 3.1 the trajectory at the top corresponds to a simulated path of the stock price process. The monotonically decreasing function depicts its running minimum. The left-hand axis is assigned to these two processes whereas the right-hand axis gives the values of the elasticities of the lookback call (grey line) and of an ordinary call (black line) with exercise price 100 which are drawn at the bottom of the figure.

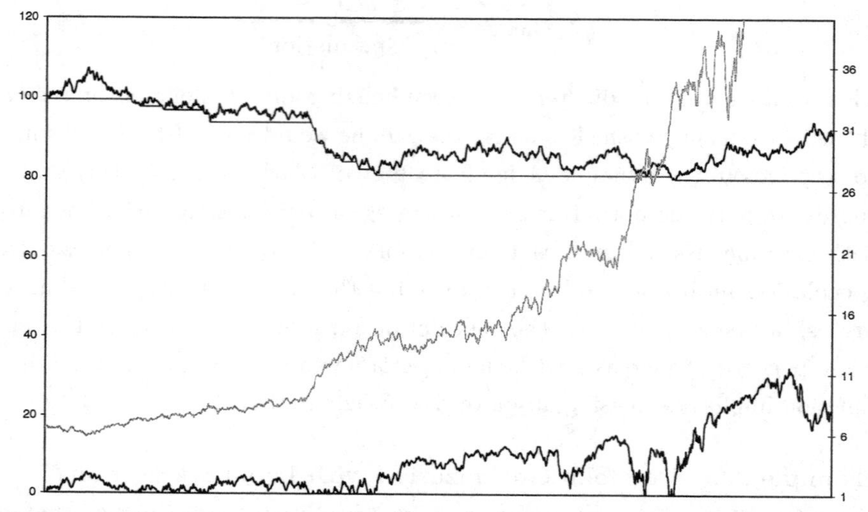

Figure 3.1: Stock price, running minimum, elasticity of a call and a lookback call

Note that the elasticity of a lookback call is defined by

$$\frac{dC_{min}/C_{min}}{dS/S} := (C_{min})_s \cdot \frac{S}{C_{min}}$$

with partial derivative $(C_{min})_s = 2 \cdot \mathcal{N}(d_1) - 1$ and

$$d_1(t) = \frac{\ln(S(t)/M(t)) + \sigma^2(T_{min} - t)}{\sigma\sqrt{T_{min} - t}}.$$

Here, $\mathcal{N}$ denotes the cumulative standard normal distribution function. For $S(t) = M(t)$ the elasticity equals one, which can be seen in Figure 3.1. The elasticity of the lookback call is clearly smaller than the elasticity of the ordinary call. This follows from the fact that the lookback call is almost always in the money. Therefore, an explosion of elasticity is very rare in contrast to an ordinary call. As detailed in Figure 3.1, the elasticity of the ordinary call explodes because it ends out of the money.

Assume now that an investor has an initial wealth of $x_0 = 100$ and maximizes utility from terminal wealth at time $T = 0.5$ with respect to the power utility

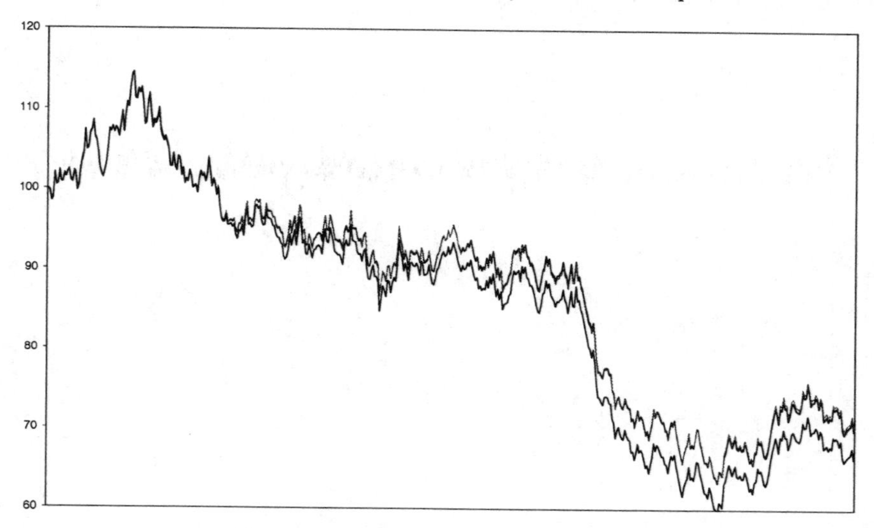

Figure 3.2: Optimal wealth processes

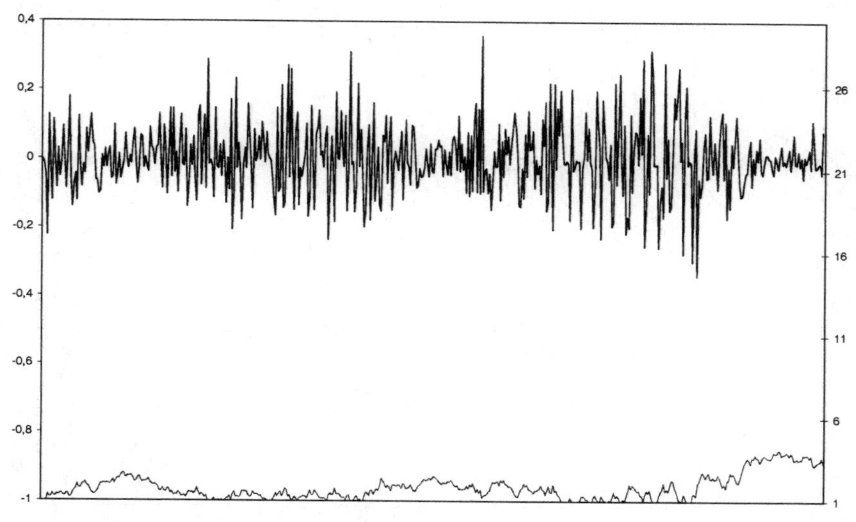

Figure 3.3: Lookback call: Percentage of change vs. elasticity

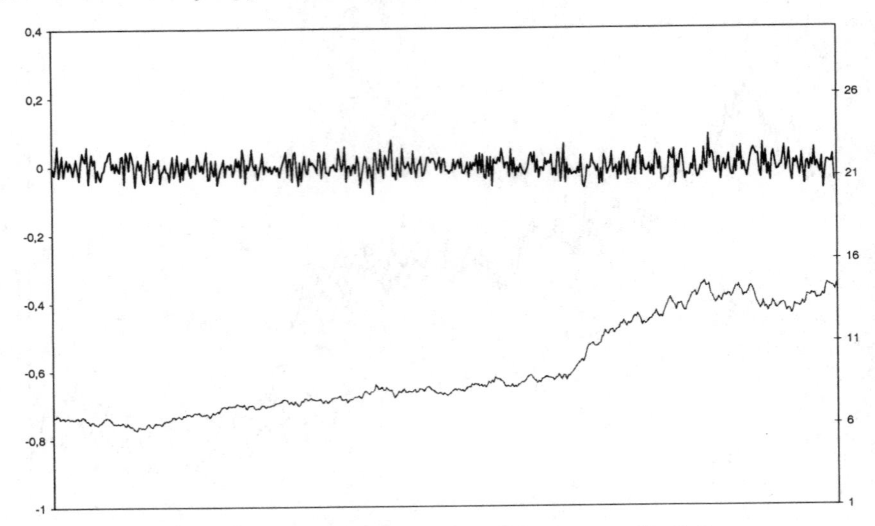

Figure 3.4: Call: Percentage of change vs. elasticity

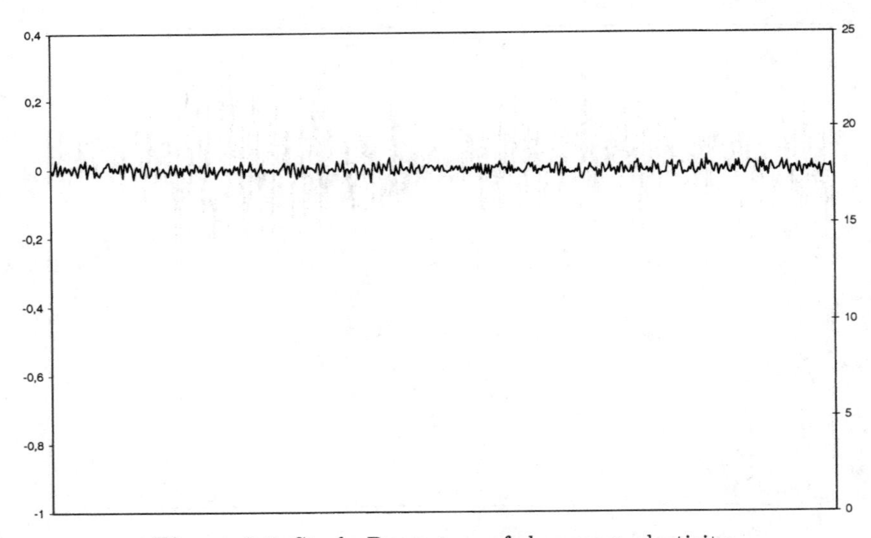

Figure 3.5: Stock: Percentage of change vs. elasticity

function $U(x) = 0.5\sqrt{x}$. Then the optimal number of lookback calls equals

$$\varphi^*_{min}(t) = \frac{1}{1-\gamma} \cdot \frac{\lambda}{\sigma^2} \cdot \frac{X(t)}{(C_{min}(t))_s S(t)} = 2 \cdot \frac{X(t)}{(C_{min}(t))_s S(t)}.$$

For an ordinary call one has to replace $(C_{min})_s$ by C_s. In Figure 3.2 we plotted the optimal wealth processes of the portfolio problems where the investor can put his funds into

- the money market account and the stock (upper trajectory),
- the money market account and the call (middle trajectory), or
- the money market account and the lookback call (lower trajectory).

From a theoretical point of view the three processes should coincide because continuous trading is assumed to take place. In our example we have assumed that the investor trades at thousand equidistant points (and not continuously) during the investment period $[0, 0.5]$. Therefore, differences between the processes can occur.

Figure 3.2 shows that it is difficult to track the optimal wealth process if the investor invests in lookback calls and if, as in our example, the stock price falls. This is so because the elasticity of a lookback call falls sharply to one if the running minimum is reached. By contrast, the wealth processes seem to be almost indistinguishable independent of whether the investor can put money into ordinary calls or into stocks. In Figure 3.3, 3.4, and 3.5 we have illustrated the elasticities (grey lines) and the percentages of change in the amounts invested in the respective assets (black lines). It is clear that the percentage

$$\frac{\varphi^*_{min}(t)C_{min}(t) - \varphi^*_{min}(t-1)C_{min}(t-1)}{\varphi^*_{min}(t-1)C_{min}(t-1)}$$

corresponding to the lookback call fluctuates most extensively. The variables $t-1$ and t are two successive time instants used to discretizate the investment period $[0, 0.5]$. The other percentages are defined analogously. We wish to stress that a lookback call meets the Black-Scholes PDE almost surely, i.e. with probability one, except for those points where the stock price touches its running minimum. Nevertheless, this last property provides significant deviations if trading occurs only at discrete points in time (and not continuously).

3.3 Duration in Portfolio Optimization

In the previous section we have explained how Merton's portfolio problem might be solved by optimizing over elasticities with respect to stocks. However, Merton's assumption of deterministic interest rates is not a realistic one, at least for long planning periods. Therefore, we relax this assumption by considering special types of affine term structures.[2] In the first place we must however demonstrate how our concept of elasticities can be applied to bond portfolio problems.

For ease of exposition, consider an affine interest rate model with one underlying risk factor which is represented by the short rate. In this case, the price of a bond with maturity T (short: T-bond) has the following representation where $K(t,T)$ and $D(t,T)$ are deterministic functions of $t \in [0,T]$:

$$P(t,T,r) = e^{K(t,T)-D(t,T)r}.$$

In most cases we will omit the functional dependency of the bond price with respect to r. According to Cox/Ingersoll/Ross (1979) the sensitivity of the bond with respect to the short rate equals

$$D(t,T) = -\frac{\partial P(t,T,r)}{\partial r} \cdot \frac{1}{P(t,T,r)},$$

which we refer to as duration of the T-Bond, although the stochastic duration defined by Cox/Ingersoll/Ross (1979) equals $G^{-1}(D(t,T))$ with a function G depending on the particular term structure model. This transformation via G is due to the fact that duration is normally measured in units of time. Only in the model of Ho/Lee (1986) duration and sensitivity coincide.[3] However, for our purposes $D(t,T)$ is the relevant term to consider.

The short rate evolves according to the SDE

$$dr(t) = a(t)dt + b(t)dW(t), \tag{3.4}$$

where with a slight abuse of notation $a(t) = a(t,r(t))$ and $b(t) = b(t,r(t))$ are measurable and sufficiently integrable so that the SDE has a unique solution. For example, choosing $b = const. > 0$ the drift $a(t) = \tilde{a}(t) + b\zeta(t)$ or $a(t) = \theta(t) - \alpha r(t) + b\zeta(t)$ leads to the models of Ho/Lee (1986) or of Vasicek (1977),

[2] See e.g. Duffie/Kan (1996).

[3] See Munk (1999).

respectively. The process $\zeta(t) = \zeta(t, r(t))$ denotes the market price of risk of the bond market. Besides, $\tilde{a}$ and θ are deterministic functions of time which reflect the initial term structure. For instance, if the initial forward curve $f(0,T)$, $0 \leq T \leq T^*$, is continuously differentiable, we obtain $\tilde{a}(t) = f_T^*(0,t) + b^2 t$. Moreover, choosing $a(t) = \kappa(\theta - r(t))$, $\kappa, \theta \in I\!R$, and $b(t) = \tilde{b}\sqrt{r(t)}$, $\tilde{b} \in I\!R$, the model of Cox/Ingersoll/Ross (1985) is implied.

One of the fundamental results in pricing interest rate sensitive contingent claims is the so-called term structure equation.[4] In particular, the bond prices in our model solve the PDE

$$P_t + (a - \zeta b)P_r + 0.5b^2 P_{rr} - rP = 0 \tag{3.5}$$

with terminal condition $P(T,T,r) = 1$. On the other hand, using Ito's formula we arrive at

$$\begin{aligned} dP &= P_t dt + P_r dr + 0.5 P_{rr} d < r > \\ &= (P_t + aP_r + 0.5b^2 P_{rr})dt + bP_r dW. \end{aligned}$$

Substituting (3.5) into the latter equation results in

$$\begin{aligned} dP &= (rP + \zeta b P_r)dt + bP_r dW \\ &= P\Big[(r + \zeta b \tfrac{P_r}{P})dt + b\tfrac{P_r}{P}\, dW\Big] \\ &= P\Big[(r - \zeta b D)dt - bD dW\Big]. \end{aligned}$$

We thus obtain an SDE for the bond price which involves the duration. Since contingent claims $C(t) = C(t, r(t))$ on the short rate also meet the term structure equation (3.5) - clearly with a different terminal condition -, a similar line of argumentation leads to the following SDE

$$dC = C\Big[(r + \zeta b \underbrace{\tfrac{C_r}{C}}_{=-D_C})dt + b \underbrace{\tfrac{C_r}{C}}_{=-D_C} dW\Big], \tag{3.6}$$

where D_C denotes the duration of the derivative. Notice that in our model contingent claims on a bond can be treated as contingent claims on the short rate. For these claims we obtain

$$C_r = \frac{\partial C}{\partial r} = \frac{\partial C}{\partial P} \cdot \frac{\partial P}{\partial r} = C_P \cdot P_r.$$

[4] See Cox/Ingersoll/Ross (1985) or Duffie (1992).

Therefore, the duration of a derivative on a bond equals its elasticity ε_C with respect to the bond times the duration D of the bond, i.e.

$$D_C = -\frac{C_r}{C} = -\frac{C_P \cdot P_r}{C} = -C_P \frac{P}{C} \cdot \frac{P_r}{P} = \varepsilon_C \cdot D.$$

Consider an investor who has the opportunity to invest in the money market account which evolves according to the SDE

$$dM(t) = M(t)r(t)dt, \quad M(0) = 1,$$

and two bonds with price dynamics given by

$$dP(t,T_1) = P(t,T_1)\Big[(r(t) + \zeta(t)\sigma_1(t))dt + \sigma_1(t)\,dW(t)\Big],$$
$$dP(t,T_2) = P(t,T_2)\Big[(r(t) + \zeta(t)\sigma_2(t))dt + \sigma_2(t)\,dW(t)\Big].$$

For instance, in the Ho-Lee model we have $\sigma_i(t) = -b(T_i - t)$, in the Vasicek model $\sigma_i(t) = \frac{b}{a}(\exp(-a(T_i - t)) - 1)$, and in both models $\sigma_i(t) = -b \cdot D_i(t)$, $i = 1, 2$. Furthermore, assume that the investor has an investment horizon T which is smaller than the maturities of the bonds. Hence, his wealth equation reads

$$dX = X\Big[(r - \zeta b(\pi_1 D_1 + \pi_2 D_2))dt - b(\pi_1 D_1 + \pi_2 D_2)dW\Big],$$

where π_1 and π_2 denote the fractions invested in the first and second bond. If we recall that the static duration D of a bond portfolio equals the weighted sum of the bond durations, i.e.

$$D = \pi_1 D_1 + \pi_2 D_2,$$

then the wealth equation can be rewritten as follows:

$$dX = X\Big[(r - \zeta bD)dt - bD\,dW\Big]. \tag{3.7}$$

From (3.6) it is apparent that, even if the investor can put money into derivatives on the short rate, his wealth equation remains the same. In analogy to the results in Section 3.2 we shall interpret the portfolio duration D as the control variable. Consequently, the optimal duration does not depend on a specific contingent claim on the short rate provided that the financial market consists of at least one claim on the short rate.

If the investor puts his funds into the money market account and into a bond, the wealth equation reads as follows:

$$dX = X\Big[(r + \zeta\pi\sigma)dt + \pi\sigma dW\Big], \tag{3.8}$$

where π denotes the percentage invested in the bond and σ stands for the volatility of the bond. Comparing both wealth equations (3.7) and (3.8), we conclude that our portfolio problem can be treated as if it were a simplified portfolio problem with one single bond. It is therefore sufficient to solve this *reduced bond portfolio problem.* The optimal duration is then given by

$$D^*(t) = \frac{-\pi^*(t)\sigma(t)}{b}, \tag{3.9}$$

where π^* denotes the optimal proportion of bonds in the reduced bond portfolio problem. Hence, we end up with the following proposition:

Proposition 3.2 (Two-Step-Portfolio-Selection with Stoch. Interest) *Consider an affine term structure model and a financial market which consists of a money market account and contingent claims on the short rate (3.4) as for example bonds. We assume that at any point in time up to the planning horizon the investor can put wealth into the money market account and into one of these claims. Then the following results are valid:*

(i) The optimal wealth process is uniquely determined by an optimal portfolio duration.

(ii) This duration can be computed without any further knowledge of the investor's opportunity set by solving a reduced bond portfolio problem and applying (3.9).

Remarks.

a) As in Proposition 3.1 the assumptions ensure that the market is complete.

b) Our result is also valid for term structure models (3.4) which do not fit into the affine term structure framework. However, in these models it is not ensured that closed form solutions for the durations of the bonds exist.

Note should be taken that the results of Proposition 3.2 can be generalized to multifactor models. The second of the following examples addresses this point.

Example 3.1: "The Models of Ho/Lee (1986) and Vasicek (1977)" Consider an investor who maximizes utility from terminal wealth at time T with respect to a power utility function $U(x) = \frac{1}{\gamma}x^\gamma$, $\gamma \in (-\infty, 0) \cup (0,1)$. Assume further that the bond market is described by the Ho-Lee model or the Vasicek model and that the market price of risk ζ is deterministic and continuous. We choose a bond with an arbitrary maturity $T_1 > T$ and volatility $\sigma(t)$. According to Section 2.2 the optimal bond proportion in the reduced problem reads[5]

$$\pi^*(t) = \underbrace{\frac{1}{1-\gamma} \cdot \frac{\zeta(t)}{\sigma(t)}}_{\text{Merton's result}} - \underbrace{\frac{\gamma}{1-\gamma} \cdot \kappa(t)}_{\text{Hedging term}} \tag{3.10}$$

with $\kappa(t) = \frac{T-t}{T_1-t}$ in the Ho-Lee model and $\kappa(t) = \frac{1-e^{-\alpha(T-t)}}{1-e^{-\alpha(T_1-t)}}$ in the Vasicek model. The first expression in (3.10) corresponds to the classical result by Merton (1969, 1971). Note that in our formulation ζ usually attains negative values, i.e. $-\zeta > 0$. The interpretation of the second term becomes clear when we apply our above results. Plugging (3.10) into (3.9) gives

$$\begin{aligned} D^*(t) &= \frac{1}{1-\gamma} \cdot \frac{-\zeta(t)}{b} + \frac{\gamma}{1-\gamma} \cdot \frac{\sigma(t)}{b} \cdot \kappa(t) \\ &= \frac{1}{1-\gamma} \cdot \frac{-\zeta(t)}{b} - \frac{\gamma}{1-\gamma} \cdot D(t,T), \end{aligned} \tag{3.11}$$

where $D(\cdot, T)$ denotes the duration of a fictitious bond with maturity equal to the investment horizon T, thus $D(t,T) = T - t$ in the Ho-Lee model and $D(t,T) = \alpha^{-1}(1 - e^{-\alpha(T-t)})$ in the Vasicek model. Hence, the second term in (3.10) can be interpreted as a hedging term. As for a very risk averse investor (i.e. $\gamma = -\infty$) the Merton term vanishes, his optimal duration corresponds to the second term. Therefore, he can achieve his optimal strategy by investing in a bond with maturity T. If we focus on the planning horizon T, this strategy provides a perfect hedge against interest rate risk resulting from the bond investment.[6]

[5] Note that in Section 2.2 we consider a power utility function $U(x) = x^\gamma$, $\gamma \in (0,1)$, which leads to the same optimal portfolio as $U(x) = \frac{1}{\gamma}x^\gamma$, $\gamma \in (0,1)$, because both utility function only differ with respect to a positive multiplicative constant. Further, the results of Proposition 2.1 and 2.2 are also valid for $U(x) = \frac{1}{\gamma}x^\gamma$ with $\gamma < 0$.

[6] See Wachter (2001) for a generalization of this result.

We wish to point out that the optimal portfolio duration does not depend on the maturity T_1 of the bond. This result is in line with Proposition 3.2 (ii) because the optimal duration has to be independent of a specific asset. Therefore, the maturity of the bond is irrelevant given that the maturity is greater than the investment horizon T. Further, note that according to Heath/Jarrow/Morton (1992) the market price of risk of the bond market is also independent of a specific bond.

Example 3.2: "Two-factor Model of Heath/Jarrow/Morton (1992)" Heath/Jarrow/Morton (1992) analyzed an example of a two-factor model. Let $[0, T^*]$ be a given trading intervall. With constants b_1, b_2, and $\alpha > 0$ they chose $(b_1, b_2 e^{-\alpha(T_f - t)})$, $0 \leq t \leq T_f \leq T^*$, as volatility surface of the forward rates so that the forward rates evolve according to the SDEs

$$df(t, T_f) = A(t, T_f)dt + b_1 dW_1(t) + b_2 e^{-\alpha(T_f - t)} dW_2(t),$$

where (W_1, W_2) denotes a two-dimensional Brownian motion and $A(t, T_f)$ is determined by the drift condition of Heath/Jarrow/Morton (1992). The first factor W_1 can be seen as a "long-run factor", whereas the second factor W_2 influences the spread between a "short" and this "long term factor".

Assuming that the market price of risk $\zeta = (\zeta_1, \zeta_2)$ follows a deterministic and continuous process, the price dynamics of a T_k-bond follow the SDE

$$\begin{aligned} dP(t, T_k) = P(t, T_k)\Big[&(r(t) + \zeta_1(t)\sigma_{k1}(t) + \zeta_2(t)\sigma_{k2}(t))dt \\ &+\sigma_{k1}(t)dW_1(t) + \sigma_{k2}(t)dW_2(t)\Big], \end{aligned} \tag{3.12}$$

$P(0, T_k) = p_0(T_k) > 0$, with $\sigma_{k1}(t) = -b_1 \cdot (T_k - t)$, $\sigma_{k2}(t) = -b_2 \cdot \alpha^{-1} \cdot (1 - e^{-\alpha(T_k - t)})$, and $T_k < T^*$.

Calculations detailed in Appendix 3.5.1 show that the short rate can be represented as the sum of the two following component processes r_1 and r_2, i.e. $r = r_1 + r_2$, which we refer to as factor rates[7]

$$\begin{aligned} dr_1(t) &= (\tilde{a}(t) + b_1 \cdot \zeta_1(t))dt + b_1 dW_1(t), \\ dr_2(t) &= (\theta(t) - \alpha r_2(t) + b_2 \cdot \zeta_2(t))dt + b_2 dW_2(t), \end{aligned}$$

where $\tilde{a}$ and θ are deterministic and continuous functions containing information about the initial term structure. The reason for this result is the additivity

[7] See Duffie/Kan (1996) for related results.

of the factors in the forward rate dynamics which transfers to the short rate. Due to this property the price of a T_k-bond can be rewritten as

$$P(t,T_k) = \exp\Big(K_1(t,T_k) - D_1(t,T_k)r_1(t)\Big)\cdot\exp\Big(K_2(t,T_k) - D_2(t,T_k)r_2(t)\Big) \tag{3.13}$$

with $D_1(t,T_k) = T_k - t$, $D_2(t,T_k) = \alpha^{-1}(1-e^{-\alpha(T_k-t)})$, and deterministic functions K_1, K_2 which we need not specify for this example. According to Heath/Jarrow/Morton (1992) the functions $D_1(\cdot,T_k)$ and $D_2(\cdot,T_k)$ should be interpreted as sensitivities to a long term factor and the spread between the long and a short term factor, respectively. Hence, we refer to them as factor durations. Plugging them into (3.12) leads to the following SDE:

$$\begin{aligned} dP(t,T_k) = P(t,T_k)\Big[&(r(t) - \zeta_1(t)b_1D_1(t,T_k) - \zeta_2(t)b_2D_2(t,T_k))dt \\ &-b_1D_1(t,T_k)dW_1(t) - b_2D_2(t,T_k)dW_2(t)\Big]. \end{aligned}$$

The wealth equation of a portfolio problem with at least two bonds having maturities greater than the investment horizon reads as follows:

$$\begin{aligned} dX(t) = X(t)\Big[&\big(r(t) - \zeta_1(t)b_1D_1(t) - \zeta_2(t)b_2D_2(t)\big)dt \\ &-b_1D_1(t)dW_1(t) - b_2D_2(t)dW_2(t)\Big], \qquad X(0) = x_0 > 0, \end{aligned}$$

where $D_1(t)$ and $D_2(t)$ denote the factor durations of the portfolio. As discussed above, they are equal to weighted sums of factor durations of the bonds or bond derivatives under consideration. Thus, the optimization problem is independent of a specific bond and one can proceed to solve the reduced problem which is formulated with respect to the factor durations. Assuming an investor who maximizes utility from terminal wealth at time T with respect to $U(x) = \frac{1}{\gamma}x^\gamma$, the optimal factor durations are given by[8]

$$D_j^*(t) = \frac{1}{1-\gamma}\cdot\frac{-\zeta_j(t)}{b_j} - \frac{\gamma}{1-\gamma}\cdot D_j(t,T), \tag{3.14}$$

$j = 1,2$, where $D_j(\cdot,T)$ denotes the j-th factor duration of a fictitious bond with maturity equal to the investment horizon T, thus $D_1(t,T) = T - t$ and $D_2(t,T) = \alpha^{-1}(1-e^{-\alpha(T-t)})$. Note that, as mentioned above, we usually have $\zeta_j < 0$, $j = 1,2$.

[8] This is a generalization of a result by Korn/Kraft (2001). The calculations are detailed in Appendix 3.5.2.

Bearing in mind our earlier findings, optimizing a bond portfolio in the two-factor model of Heath/Jarrow/Morton (1992) is in last analysis reduced to solving the following system of linear equations with respect to $\pi(t)$

$$-\text{diag}(b)D^*(t) = \sigma(t)'\pi(t), \tag{3.15}$$

where $D^* = (D_1^*, D_2^*)$ and diag(b) denotes a diagonal matrix with b_1 and b_2 on its diagonal. Given a number of d bonds $\sigma(t)$ is a $d \times 2$-matrix with transpose $\sigma(t)'$ and $\pi(t)$ corresponds to a vector with d components.

For the solvability of (3.15) it will be an important issue whether the matrix $\sigma(t)$ is of maximum rank for all $t \in [0, T]$. Fortunately, as shown in Appendix 3.5.3, given two arbitrary bonds with maturities $T_1, T_2 > t$ this property is satisfied. This is not self-evident as there exists at least one obvious example - a two-factor Ho-Lee model - for which this does not apply. There, the factor durations have to coincide so that the investor can choose the durations only simultaneously. Clearly, this example is not an important one because a two-factor Ho-Lee model is questionable in itself.

Hence, an investor is able to match the optimal strategy D^* if at any time instant $t \in [0, T]$ he can trade in at least two different bonds. Note that these do not necessarily have to be the same two bonds because due to Proposition 3.2 his optimal wealth process is independent of specific bonds. As an aside, it may be pointed out that coupon bonds can be easily incorporated in our model since a coupon bond corresponds to a portfolio of (zero) bonds. Consequently, its duration equals the weighted sum of the durations of these (zero) bonds.

Example 3.3: "An Investor without Investment Opportunities"

Consider an investment problem where the investor can trade in at least one bond or can put funds into the money market account. Denote the greatest maturity of the bonds by T_B. In contrast to the previous example we assume that his investment horizon T is greater than the maturity T_B, i.e. $T > T_B$. Consequently, the investor is forced to put his wealth into the money market account when the T_B-bond will have been redeemed. Given that the investor maximizes utility from terminal wealth with respect to a power utility function $U(x) = \frac{1}{\gamma}x^\gamma$, the value function of the portfolio problem reads

$$v(t,x,r) = \sup_{D \in \mathcal{A}(t,x),\ D|_{[T_B,T]} \equiv 0} \mathrm{E}^{t,x,r}\left(\tfrac{1}{\gamma} X^D(T)^\gamma\right),$$

where $\mathcal{A}(t,x)$ denotes the set of all admissible durations corresponding to the initial condition (t,x).[9] From time T_B onward only the duration of the money market account is attainable for the investor which means $D(t) = 0$, $t \geq T_B$. Hence, we can explicitly calculate the value function by omitting the supremum and computing expectations. Given $t \geq T_B$ this leads to

$$\begin{aligned} v(t,x,r) &= \mathrm{E}^{t,x,r}\left(\tfrac{1}{\gamma}X(T)^\gamma\right) \\ &= \mathrm{E}^{t,x,r}\left(\tfrac{1}{\gamma}x^\gamma \exp\left(\gamma \int_t^T r(s)\,ds\right)\right) \\ &= \tfrac{1}{\gamma}x^\gamma \mathrm{E}^{t,x,r}\left(\exp\left(\gamma \int_t^T r(s)\,ds\right)\right). \end{aligned}$$

Given the Vasicek model, by Proposition 1.1, the SDE for the short rate has the following solution

$$r(t) = e^{-\alpha t}\left(r_0 + \int_0^t e^{\alpha u}\Big(\theta(u) + b\zeta(u)\Big)\,du + \int_0^t be^{\alpha u}\,dW(u)\right).$$

Therefore, we obtain[10]

$$\begin{aligned} \int_t^T r(s)\,ds = \alpha^{-1}\Big(1 - e^{-\alpha(T-t)}\Big)r_t + \underbrace{\int_t^T \int_t^s e^{\alpha(u-s)}\Big(\theta(u) + b\zeta(u)\Big)\,du\,ds}_{=:A_1(t,T)} \\ + \int_t^T \underbrace{\int_u^T be^{\alpha(u-s)}\,ds}_{=:A_2(u,T)}\,dW(u) \end{aligned}$$

Let $A_3(t,T) := \int_t^T A_2(u,T)^2\,du$ be the variance of the Ito integral. Then, we arrive at

$$\begin{aligned} &\mathrm{E}^{t,x,r}\left(\exp\left(\gamma \int_t^T r(s)\,ds\right)\right) = \\ &\qquad \underbrace{\exp\Big(\gamma A_1(t,T) + 0.5\gamma^2 A_3(t,T)\Big)}_{=:K_1(t)} \cdot \underbrace{\exp\Big(\gamma\alpha^{-1}(1 - e^{-\alpha(T-t)})r\Big)}_{=:K_2(t,r)}, \end{aligned}$$

which gives us an explicit representation of the value function for $t \geq T_B$

[9] Note that given an admissible duration $D \in \mathcal{A}(0,x)$ the killed process $D1_{[0,T_B)}$ is also admissible. Here, 1_E denotes the indicator function for E.

[10] See Ikeda/Watanabe (1981, pp. 117ff) for the interchange of Lebesgue and Ito integrals.

$$v(t,x,r) = \tfrac{1}{\gamma}x^{\gamma} \cdot K_1(t) \cdot K_2(t,r).$$

Choosing $t = T_B$ leads to an alternative terminal condition for the optimization problem up to time T_B. Hence, the value function up to time T_B can be calculated from the Hamilton-Jacobi-Bellman equation

$$\sup_D \left\{ v_t + 0.5(x^2 b^2 D^2 v_{xx} - 2xb^2 D v_{xr} + b^2 v_{rr}) + x(r - bD\zeta)v_x + av_r \right\} = 0 \tag{3.16}$$

with $v(T_B, x, r) = \frac{1}{\gamma}x^{\gamma} \cdot K_1(T_B) \cdot K_2(T_B, r)$. As shown in Appendix 3.5.4 one can solve for the value function, which reads

$$v(t,x,r) = \tfrac{1}{\gamma}x^{\gamma} K_1(T_B) \exp\left(\tfrac{1}{1-\gamma}(H(t) - H(T_B)) + \underbrace{\gamma\alpha^{-1}(1 - e^{-\alpha(T-t)})}_{=:\beta(t)} r \right) \tag{3.17}$$

for $0 \le t \le T_B$. Here, H stands for a continuously differentiable function whose exact representation is irrelevant to our following considerations. The optimal duration up to time T_B is given by

$$\begin{aligned} D^*(t) &= \frac{\zeta(t)}{b} \frac{v_x(t,x,r)}{xv_{xx}(t,x,r)} + \frac{v_{xr}(t,x,r)}{xv_{xx}(t,x,r)} \\ &= \frac{1}{1-\gamma} \cdot \frac{-\zeta(t)}{b} - \frac{\gamma}{1-\gamma} \cdot D(t,T) \end{aligned}$$

and coincides with solution (3.11) because β does not involve the maturity T_B but the investment horizon T. Hence, if the investment horizon of an investor is larger than the maturities of the traded bonds, he should ignore this restriction until it becomes restrictive. But, in reality, this is unlikely to happen because new bonds will be issued. This *invariance property of the optimal duration* is a convenient result for investors with long planning horizons such as young adults who put money aside for their retirements. It should be stressed that this property cannot be proved easily if the portfolio process π is chosen to be the control variable, the reason being that

$$\lim_{t \nearrow T_B} \pi^*_{T_B}(t) = \infty$$

for the optimal percentage $\pi^*_{T_B}$ invested in the bond with maturity T_B. Hence, in contrast to the optimal duration the optimal portfolio process is unbounded. From a technical point of view this unboundedness complicates the verification of the result.

3.4 Conclusion

In this chapter we have developed the elasticity approach to portfolio optimization (EAPO). As part of our analysis, we have examined Merton's portfolio problem and shown that the appropriate control variables are the elasticities to the traded stocks. If bond portfolio problems with stochastic interest rates are considered, the proper control variables are the factor durations. Applying this approach to portfolio optimization simplifies portfolio problems considerably because it is sufficient to solve reduced portfolio problems in a first step. For example, if the dynamics of the yield curve can be described by a two-factor Heath-Jarrow-Morton model and an investor puts his wealth into twenty bonds, the number of control variables decreases from twenty (proportions) to two (factor durations). Having computed the optimal elasticities or factor durations, in a second step, it is straightforward to determine a portfolio strategy which tracks the optimal elasticities or factor durations, respectively. Consequently, as the martingale approach, the EAPO is a kind of two-step procedure. Nevertheless, it need to be stressed that the EAPO works in both Merton's optimal control approach and the martingale approach. Another important feature of the EAPO is that non-linear contracts on traded assets such as options can be easily included in portfolio problems. Therefore, in Chapter 5 we will apply this approach to solve portfolio problems with defaultable securities in a firm value framework, the reason being that, as an example, defaultable bonds can be modeled as contingent claims on firm value. Moreover, we will address an important restriction which occurs in firm value frameworks and leads to an incomplete market. Nevertheless, the EAPO is still applicable.

3.5 Appendix

3.5.1 Representation of the Short Rate by Factor Rates

Let $T^* > 0$ and consider a two-factor Heath-Jarrow-Morton model given by the forward rate dynamics

$$df(t,T) = A(t,T)dt + B(t,T)dW(t)$$

with initial term structure $f(0,T) = f^*(0,T)$, $0 \le t \le T \le T^*$, and with $B(t,T) = (b_1, b_2 e^{-\alpha(T-t)})'$, $0 \le t \le T \le T^*$. Note that in this subsection the variable T does not stand for the investment horizon but for the maturity of the forward rate $f(t,T)$. The Heath-Jarrow-Morton drift condition yields

$$A(t,T) = b_1^2(T-t) + \tfrac{b_2^2}{\alpha} e^{-\alpha(T-t)}(1 - e^{-\alpha(T-t)}) + \sum_{j=1}^{2} B_j(t,T)\zeta_j(t).$$

Hence, we get

$$\begin{aligned} df(t,T) = {} & \Big[b_1^2(T-t) + \tfrac{b_2^2}{\alpha} e^{-\alpha(T-t)}(1 - e^{-\alpha(T-t)}) + \sum_{j=1}^{2} B_j(t,T)\zeta_j(t)\Big] dt \\ & + b_1 dW_1(t) + b_2 e^{-\alpha(T-t)} dW_2(t). \end{aligned}$$

Assuming a differentiable initial term structure, we compute the derivative with respect to T

$$\begin{aligned} f_T(t,T) = {} & f_T^*(0,T) + b_1^2 t + \int_0^t \Big(b_2^2(2e^{-2\alpha(T-s)} - e^{-\alpha(T-s)}) \\ & - b_2 \alpha e^{-\alpha(T-s)}\zeta_2(s)\Big)\, ds - \int_0^t \alpha b_2 e^{-\alpha(T-s)} dW_2(s) \end{aligned}$$

and arrive at

$$\begin{aligned} f_T(t,t) &= f_T^*(0,t) + b_1^2 t - \alpha\Big[\int_0^t \tfrac{b_2^2}{\alpha}(e^{-\alpha(t-s)} - 2e^{-2\alpha(t-s)}) \\ &\quad + b_2 e^{-\alpha(t-s)}\zeta_2(s)\, ds + \int_0^t b_2 e^{-\alpha(t-s)} dW_2(s)\Big] \\ &= f_T^*(0,t) + b_1^2 t + \tfrac{b_2^2}{2\alpha}(1 - e^{-2\alpha t}) + \alpha h(t) - \alpha r_2(t), \end{aligned}$$

where

$$\begin{aligned} r_2(t) := h(t) + \int_0^t \tfrac{b_2^2}{\alpha}(e^{-\alpha(t-s)} - e^{-2\alpha(t-s)}) + b_2 e^{-\alpha(t-s)}\zeta_2(s)\, ds \qquad (3.18) \\ + \int_0^t b_2 e^{-\alpha(t-s)} dW_2(s) \end{aligned}$$

and h stands for a continuously differentiable function. In a one-factor Vasicek model h would be equal to $f^*(0,t)$.[11] In our context, we can for example choose $h(t) = 0.5f^*(0,t)$, but this is not absolute. Now there exists a continuous function θ so that the dynamics of r_2 can be represented by

$$dr_2(t) = \Big(\theta(t) - \alpha r_2(t) + b_2\zeta_2(t)\Big)dt + b_2 dW_2(t) \tag{3.19}$$

with $r_2(0) = h(0)$ and

$$\theta(t) = h'(t) + \alpha h(t) + \tfrac{b_2^2}{2\alpha}(1 - e^{-2\alpha t}). \tag{3.20}$$

To prove this result one can compare the solution

$$r_2(t) = e^{-\alpha t}\left(r_2(0) + \int_0^t e^{\alpha u}\Big(\theta(u) + b_2\zeta_2(u)\Big)\,du + \int_0^t b_2 e^{\alpha u}\,dW_2(u)\right)$$

of the SDE (3.19) with the representation (3.18) of r_2. This leads to an integral equation

$$e^{-\alpha t}r_2(0) + e^{-\alpha t}\int_0^t e^{\alpha u}\theta(u)\,du = h(t) + \tfrac{b_2^2}{2\alpha^2}(1 - 2e^{-\alpha t} + e^{-2\alpha t})$$

for θ. Differentiating with respect to t gives

$$-\alpha e^{-\alpha t}r_2(0) - \alpha e^{-\alpha t}\int_0^t e^{\alpha u}\theta(u)\,du + \theta(t) = h'(t) + \tfrac{b_2^2}{\alpha}e^{-\alpha t}(1 - e^{-\alpha t}).$$

Hence, the integral can be eliminated and we get (3.20). This leads to the following dynamics of the short rate[12]

$$\begin{aligned} dr(t) &= \Big(f_T(t,t) + A(t,t)\Big)dt + B_1(t,t)dW_1(t) + B_2(t,t)dW_2(t) \\ &= \Big(f_T^*(0,t) + b_1^2 t + \tfrac{b_2^2}{2\alpha}(1 - e^{-2\alpha t}) + \alpha h(t) - \alpha r_2(t) \\ &\quad + b_1\zeta_1(t) + b_2\zeta_2(t)\Big)dt + b_1 dW_1(t) + b_2 dW_2(t) \end{aligned}$$

with $r(0) = f^*(0,0)$. Let $r_1 := r - r_2$. Then we obtain

$$dr_1(t) = \Big[\underbrace{f_T^*(0,t) + b_1^2 t - h'(t)}_{=:\tilde{a}(t)} + b_1\zeta_1(t)\Big]dt + b_1\,dW_1(t)$$

with $r_1(0) = r(0) - r_2(0) = f^*(0,0) - h(0)$. The bond price dynamics are given by

[11] See e.g. Musiela/Rutkowski (1997), pp 323f.

[12] See e.g. Björk (1998), pp. 232ff.

$$P(t,T) = P(0,T)\exp\Big(\int_0^t r(s) + \tilde{A}(s,T)\,ds + \int_0^t \tilde{B}(s,T)\,dW(s)\Big),$$

where

$$\tilde{A}(t,T) = -\int_t^T A(t,s)\,ds, \qquad \tilde{B}(t,T) = -\int_t^T B(t,s)\,ds.$$

Therefore, the additive structures of the forward rate and the short rate dynamics provide a product representation of each bond price where one factor equals a fictitious bond price of the Vasicek model and the other factor equals a fictitious bond price of the Ho-Lee model. Since both models possess an affine term structure, there exists a representation (3.13) of the bond price

$$P(t,T)\exp\Big(K_1(t,T) - D_1(t,T)r_1(t)\Big)\cdot\exp\Big(K_2(t,T) - D_2(t,T)r_2(t)\Big)$$

with $D_1(t,T) = T - t$, $D_2(t,T) = \alpha^{-1}(1 - e^{-\alpha(T-t)})$, and deterministic functions K_1 and K_2. Since

$$-\frac{\partial P(t,T)}{\partial r_j(t)}\cdot\frac{1}{P(t,T)} = D_j(t,T),$$

we can interpret the process D_j as the duration with respect to factor j. Therefore, we have $\sigma_{kj} = -b_j D_j(\cdot, T_k)$, $j = 1, 2$. Note that these factor durations are a generalization of the ideas by Cox/Ingersoll/Ross (1979). In addition, we wish to stress that the results of this subsection can be generalized in a straightforward manner to a multifactor version of the discussed model.

3.5.2 Optimal Factor Durations (3.14)

Define $a = (a_1, a_2)' = (\tilde{a} + b_1\zeta_1, \theta - \alpha r_2 + b_2\zeta_2)'$ and $b = (b_1, b_2)$. We face a portfolio problem with a three-dimensional state process (X, r_1, r_2). The Hamilton-Jacobi-Bellman equation (HJB) reads as follows:

$$\sup_D \Big(0,5x^2 D'\mathrm{diag}(b^2)DG_{xx} - xD'\mathrm{diag}(b^2)G_{xr} + 0,5b'\mathrm{diag}(G_{rr})b$$
$$+G_t + xG_x(r - D'\mathrm{diag}(b)\zeta) + a'G_r\Big) = 0$$

with terminal condition $G(T, x, r_1, r_2) = \frac{1}{\gamma}x^\gamma$, where $D = (D_1, D_2)'$, $G_r = (G_{r_1}, G_{r_2})'$, $b^2 = (b_1^2, b_2^2)$, and $G_{xr} = (G_{xr_1}, G_{xr_2})'$. Besides, $\mathrm{diag}(G_{rr})$ ($\mathrm{diag}(b^2)$) denotes a diagonal matrix with $G_{r_1r_1}$ and $G_{r_2r_2}$ (with b_1^2 and b_2^2) on its diagonal.

Assuming $G_{xx} < 0$ we get

$$xG_{xx}\mathrm{diag}(b^2)D = G_x\mathrm{diag}(b)\zeta + \mathrm{diag}(b^2)G_{xr}.$$

For $x > 0$ and $b_j \neq 0$, $j = 1, 2$, this system of equations has a unique solution

$$D^* = \mathrm{diag}(1/b)\zeta\frac{G_x}{xG_{xx}} + \frac{G_{xr}}{xG_{xx}}. \tag{3.21}$$

Plugging this solution into the HJB leads to a partial differential equation (PDE) for G

$$\begin{aligned} 0 = G_tG_{xx} + xrG_xG_{xx} - 0,5\zeta'\zeta G_x^2 - G_xG_{xr}'\mathrm{diag}(b)\zeta + a'G_rG_{xx} \\ -0,5G_{xr}'\mathrm{diag}(b^2)G_{xr} + 0,5b'\mathrm{diag}(G_{rr})bG_{xx} \end{aligned} \tag{3.22}$$

with $G(T, x, r_1, r_2) = \gamma^{-1}x^\gamma$. Using the ansatz $G(t, x, r_1, r_2) = x^\gamma f(t, r_1, r_2)$ with $f(T, r_1, r_2) = \gamma^{-1}$ leads to a PDE for f

$$\begin{aligned} 0 = (\gamma-1)ff_t - 0,5\gamma\sum_{j=1}^{2} b_j^2f_j^2 - 0,5\gamma\sum_{j=1}^{2}\zeta_j^2f^2 + 0,5(\gamma-1)\sum_{j=1}^{2} b_j^2ff_{jj} \\ -\gamma\sum_{j=1}^{2}\zeta_jb_jff_j + r\gamma(\gamma-1)f^2 + (\gamma-1)\sum_{j=1}^{2} a_jff_j, \end{aligned} \tag{3.23}$$

where $f_j = f_{r_j}$ and $f_{jj} = f_{r_jr_j}$. Now we apply the ansatz $f(t, r_1, r_2) = g(t)\cdot\exp(\sum_{j=1}^2\beta_j(t)r_j)$ with $\beta_j(T) = 0$ and $g(T) = \gamma^{-1}$. This leads to a differential equation for g

$$\begin{aligned} 0 = (\gamma-1)g' + (\gamma-1)(\beta_1' + \gamma)r_1g + (\gamma-1)(\beta_2' + \gamma - \alpha\beta_2)r_2g \\ -0,5\gamma\sum_{j=1}^{2}\zeta_j^2g - 0,5\sum_{j=1}^{2} b_j^2\beta_j^2g - \gamma\sum_{j=1}^{2}\zeta_jb_j\beta_jg + (\gamma-1)\sum_{j=1}^{2}\hat{a}_j\beta_jg \end{aligned} \tag{3.24}$$

with $\hat{a} = (a_1, \theta + b_2\zeta_2)'$. For the Ho-Lee part we choose $\beta_1 = T - t$ and for the Vasicek part $\beta_2(t) = \frac{\gamma}{\alpha}(1 - e^{\alpha(t-T)})$. We thus obtain an ordinary differential equation for g

$$0 = (1-\gamma)g' + \underbrace{\Big(0,5\gamma\sum_{j=1}^{2}\zeta_j^2 + 0,5\sum_{j=1}^{2} b_j^2\beta_j^2 + \gamma\sum_{j=1}^{2}\zeta_jb_j\beta_j + (1-\gamma)\sum_{j=1}^{2}\hat{a}_j\beta_j\Big)}_{:=-h(t)}g \tag{3.25}$$

with terminal condition $g(T) = \gamma^{-1}$. Hence, we get

$$g(t) = \gamma^{-1} \exp\left(\tfrac{1}{1-\gamma}(H(t) - H(T))\right),$$

where H is a primitive of the continuous function h. Consequently, the candidate for the value function reads

$$G(t, x, r_1, r_2) = \gamma^{-1} x^{\gamma} \exp\left(\tfrac{1}{1-\gamma}(H(t) - H(T)) + \sum_{j=1}^{2} \beta_j(t) r_j\right).$$

Computing the partial derivatives and substituting them into (3.21) yields the optimal factor durations

$$D^*(t) = \frac{-1}{1-\gamma}\mathrm{diag}(1/b)\zeta - \frac{1}{1-\gamma}\beta(t) = \frac{-1}{1-\gamma}\mathrm{diag}(1/b)\zeta - \frac{\gamma}{1-\gamma} D_T(t),$$

where $D_T(t) = \gamma^{-1}\beta(t) = (T - t, \alpha^{-1}(1 - e^{-\alpha(T-t)}))'$. The verification of this result, i.e. the proof that G is indeed the value function of the problem, can be carried out as in Section 2.2.

3.5.3 Solution of the System of Linear Equations (3.15)

For $d = 2$ consider the system of linear equations (3.15)

$$\begin{aligned} -b_1 D_1^*(t) &= -b_1(T_1 - t) \quad \pi_1(t) - \quad b_1(T_2 - t) \quad \pi_2(t), \\ -b_2 D_2^*(t) &= -\tfrac{b_2}{\alpha}(1 - e^{-\alpha(T_1 - t)})\, \pi_1(t) - \tfrac{b_2}{\alpha}(1 - e^{-\alpha(T_2 - t)})\, \pi_2(t). \end{aligned}$$

These equations have a unique solution for $0 \le t \le T_1 < T_2$ if

$$(T_1 - t)(1 - e^{-\alpha(T_2 - t)}) - (T_2 - t)(1 - e^{-\alpha(T_1 - t)}) \neq 0$$

for all $0 \le t \le T_1 < T_2$. For the sake of simplicity, let $x := T_1 - t$ and $y := T_2 - t$ so that $0 \le x \le T_1$ and $x < y$. Assume now that there exist x, y with

$$x(1 - e^{-\alpha y}) - y(1 - e^{-\alpha x}) = 0.$$

With the definition $f(z) = e^{-\alpha z}$ this implies

$$\frac{f(y) - f(0)}{y - 0} = \frac{f(x) - f(0)}{x - 0}.$$

Since f is strictly monotonically decreasing and convex, we conclude that $x = y$, which contradicts the assumption $x < y$. Consequently, the system of linear equations (3.15) has a unique solution for all $t \in [0, T_1)$.

3.5.4 Calculation of the Value Function (3.17)

Obviously, the value function meets the PDE (3.22) where each parameter with index 1 is zero and $r = r_2$. But in this example we face the different terminal condition $G(T_B, x, r) = \frac{1}{\gamma} x^\gamma K_1(T_B) K_2(T_B, r)$, which depends on the short rate r. Nevertheless, the separations of Appendix 3.5.2 can be applied. First, the ansatz $G(t, x, r) = x^\gamma f(t, r)$ with $f(T_B, r) = \gamma^{-1} K_1(T_B) K_2(T_B, r)$ leads to the PDE (3.23) for f. Considering $f(t, r) = g(t) \exp(\beta(t) r)$ with terminal conditions $g(T_B) = \gamma^{-1} K_1(T_B)$ and

$$\beta(T_B) = \alpha^{-1} \gamma (1 - e^{\alpha(T_B - T)}) \tag{3.26}$$

results in the PDE (3.24). The critical point here is that the choice $\beta(t) = \alpha^{-1} \gamma (1 - e^{\alpha(t-T)})$ does not violate condition (3.26). Hence, we get the ordinary differential equation (3.25) for g with terminal condition $g(T_B) = \gamma^{-1} K_1(T_B)$. Consequently, we arrive at

$$g(t) = \gamma^{-1} K_1(T_B) \exp\left(\tfrac{1}{1-\gamma} (H(t) - H(T_B)) \right).$$

Therefore, we obtain (3.17).

4

Barrier Derivatives with Curved Boundaries

4.1 Introduction

Barrier options are subject to intensive research. Merton (1973) was the first who derived a closed-form solution for the down-and-out call with constant barrier. Further references are Cox/Rubinstein (1985), Rubinstein/Reiner (1991), and Carr (1995). For options with lower and upper barriers closed-form solutions are not available, but option prices can be represented by infinite series. This was shown by Kunitomo/Ikeda (1992) in the more general case of deterministic exponential boundaries. More recently, Rogers/Zane (1997), Novikov/Frishling/Kordzakhia (1999), and Kolkiewicz (2002) considered related problems by different methods. All authors assumed deterministic interest rates.

Björk (1998, pp. 185f) proved a general formula for the pricing of *arbitrary* European barrier derivatives, but he also assumed constant interest rates and a constant boundary. In this chapter we relax these two assumptions. First, we consider a claim subject to a deterministic exponential boundary. We then look at the problem assuming a Gaussian interest rate model and a discounted boundary. This special form of the boundary is of particular interest because it leads to the convenient property that, for example, the down-and-out call can be statically hedged in terms of Bowie/Carr (1994) if its strike is greater than its barrier. Since each in-contract can be represented as the difference between the contract without barrier and the corresponding out-contract - this property is also known as the "in-out-parity" -, we only deduce the results for out-contracts.

This chapter is organized as follows. Since we apply some of Björk's arguments to our problem, Section 4.2 gives a short survey of those results which are relevant to our considerations. In Section 4.3 barrier derivatives with deterministic exponential boundaries are analyzed under deterministic interest rates. Assuming a Gaussian interest rate model, barrier derivatives with discounted boundaries are considered in Section 4.4. In Section 4.5 we apply this result to determine the value of debt in the firm value framework of Briys/de Varenne (1997). Section 4.6 concludes.

4.2 Björk's Result

Let us first lay down some basic notations. Given a stochastic process X on $[0, T]$ and a real number $y \in I\!R$ the hitting time of y is defined by

$$\tau(y) := \tau := \inf\{t \geq 0 | X(t) = y\}.$$

The X-process absorbed at y is given by

$$X_y(t) = X(\min\{t, \tau\}).$$

Besides, for a function Φ and a constant c the functions Φ_c and Φ^c are defined by

$$\Phi_c(x) = \begin{cases} \Phi(x), \; x > c \\ 0, \quad x \leq c \end{cases} \qquad \text{and} \qquad \Phi^c(x) = \begin{cases} \Phi(x), \; x < c \\ 0, \quad x \geq c. \end{cases}$$

For the Black-Scholes model it is well-known that under the risk-neutral measure Q the stock price follows the stochastic differential equation

$$dS(t) = S(t)\Big[rdt + \sigma dW(t)\Big], \qquad S(0) = s_0,$$

where the constants r and σ denote the riskless interest rate and the stock price volatility. The process W is a Brownian motion. We consider contingent claims of the form

$$\mathcal{Z} = \Phi(S(T)),$$

where Φ stands for a measurable function, the so-called *contract function* of the claim. The *pricing function* of $\mathcal{Z}$ is denoted by $F(t, s; T, \Phi)$, i.e. the Black-Scholes price of $\mathcal{Z}$ at time t is $F(t, s; T, \Phi)$, given that $S(t) = s$. Such contracts are called T-contracts. In most cases we omit the variable T. The corresponding down-and-out-T-contract is defined by

$$\mathcal{Z}_{LO} = \begin{cases} \Phi(S(T)), & S(t) > L(t) \text{ for all } t \in [0,T], \\ 0, & S(t) \leq L(t) \text{ for some } t \in [0,T], \end{cases}$$

where $L(t)$ stands for the barrier. Further, $F_{LO}(t,s;T,\Phi)$ denotes the pricing function. Assuming a *constant barrier* $L(t) = L$ Björk (1998, p. 185) proved the following proposition:

Proposition 4.1 (Pricing Down-and-out Contracts) *If we consider a T-claim $\mathcal{Z} = \Phi(S(T))$, then the pricing function of the corresponding down-and-out contract $\mathcal{Z}_{LO}$ is given, for $s > L$, by*

$$F_{LO}(t,s;\Phi) = F(t,s;\Phi_L) - \left(\frac{L}{s}\right)^{\frac{2r}{\sigma^2}-1} F\left(t, \frac{L^2}{s}; \Phi_L\right).$$

Proof. We sketch Björk's proof. Without loss of generality we set $t = 0$. Assume that $S(0) = s > L$, and recall that S_L denotes the stochastic process S with (possible) absorption at L, whereas Φ_L denotes the real function $\Phi_L(x) := \Phi(x) \cdot 1_{\{x>L\}}$. To shorten notations, we write inf instead of $\inf_{0 \leq t \leq T}$. Using the risk neutral valuation principle, we arrive at

$$\begin{aligned} F_{LO}(0,s;\Phi) &= e^{-rT} \mathrm{E}_Q^{0,s}(\mathcal{Z}_{LO}) = e^{-rT} \mathrm{E}_Q^{0,s}\Big(\Phi(S(T)) \cdot 1_{\{\inf S(t) > L\}}\Big) \\ &= e^{-rT} \mathrm{E}_Q^{0,s}\Big(\Phi_L(S_L(T)) \cdot 1_{\{\inf S(t) > L\}}\Big) = e^{-rT} \mathrm{E}_Q^{0,s}\Big(\Phi_L(S_L(T))\Big). \end{aligned}$$

Now Björk shows that

$$\mathrm{E}_Q^{0,s}(\Phi_L(S_L(T))) = \mathrm{E}_Q^{0,s}\Big(\Phi_L(S(T))\Big) - \left(\frac{L}{s}\right)^{\frac{2r}{\sigma^2}-1} \mathrm{E}_Q^{0,\frac{L^2}{s}}\Big(\Phi_L(S(T))\Big).$$

We refer the reader to Björk's book for further details. □

Basically, Proposition 4.1 says that computing the price of a down-and-out derivative reduces to the standard problem of computing the price of the ordinary related claim without barrier. The key for this result is the existence of an explicit formula for the transition density of Brownian motion with drift.

A similar proposition for up-and-out contracts is given by Björk (1998, p. 188). To formulate his result, we define an up-and-out-T-contract by

$$\mathcal{Z}^{LO} = \begin{cases} \Phi(S(T)), \; S(t) < L(t) \text{ for all } t \in [0,T], \\ \quad 0, \qquad S(t) \geq L(t) \text{ for some } t \in [0,T], \end{cases}$$

where $L(t)$ stands for the barrier of the claim. Denoting the pricing function by $F^{LO}(t,s;T,\Phi)$, Björk proved the following proposition:

Proposition 4.2 (Pricing Up-and-out Contracts) *If we consider a T-claim $\mathcal{Z} = \Phi(S(T))$, then the pricing function of the corresponding up-and-out contract $\mathcal{Z}^{LO}$ is given, for $s < L$, by*

$$F^{LO}(t,s;\Phi) = F(t,s;\Phi^L) - \left(\frac{L}{s}\right)^{\frac{2r}{\sigma^2}-1} F\left(t, \frac{L^2}{s}; \Phi^L\right)$$

Proof. Similar arguments apply as in the proof of Proposition 4.1. □

4.3 Deterministic Exponential Boundaries

First, let $F(t,s;b,\Phi)$ be the pricing function of a T-claim $\Phi(S(T))$ if the cost-of-carry is b. In case of a stock paying the continuous dividend yield y we obtain $b = r - y$. If $y = 0$, we omit at times the variable b. Note that we only introduce this notation for derivatives without barrier. Now assume an *exponential barrier* given by

$$L(t) = A \cdot e^{\delta t} \tag{4.1}$$

with $A > 0$, $\delta \in \mathbb{R}$. Then the following proposition for down-and-out contracts is valid:

Proposition 4.3 (Pricing Down-and-out Contracts) *Assume an exponential barrier (4.1). If we consider a T-claim $\mathcal{Z} = \Phi(S(T))$, then the pricing function of the corresponding down-and-out contract $\mathcal{Z}_{LO}$ is given, for $s > L(t)$, by*

$$F_{LO}(t,s;\Phi) = F(t,\tilde{s};b,\Phi_{\tilde{L}}) - \left(\frac{\tilde{L}}{\tilde{s}}\right)^{\frac{2b}{\sigma^2}-1} F\left(t, \frac{\tilde{L}^2}{\tilde{s}}; b, \Phi_{\tilde{L}}\right), \tag{4.2}$$

where $\tilde{s} := s \cdot e^{\delta(T-t)}$, $\tilde{L} := A \cdot e^{\delta T}$, and $b := r - \delta$. The price given by the pricing function $F(\cdot,\cdot;b,\cdot)$ has to be calculated as if the drift of the stock were equal to b.

Proof. Without loss of generality we set $t = 0$ and choose $s = s_0 = S(0) > L(0)$. Again, to shorten notations, we write inf instead of $\inf_{0 \le t \le T}$. Using the risk neutral valuation principle, we have

$$F_{LO}(0, s; \Phi) = e^{-rT} \mathrm{E}_Q^{0,s}(\mathcal{Z}_{LO}) = e^{-rT} \mathrm{E}_Q^{0,s}\Big(\Phi(S(T)) \cdot 1_{\{S(t) > L(t) \ \forall \ 0 \le t \le T\}}\Big).$$

Since

$$\begin{aligned}\{S(t) > L(t) \ \forall \ 0 \le t \le T\} &= \{S(t) > A \cdot e^{\delta t} \ \forall \ 0 \le t \le T\} \\ &= \{\inf S(t) e^{-\delta t} > A\},\end{aligned}$$

using the definitions $\tilde{s}(t) := S(t) e^{\delta(T-t)}$ and $\tilde{L} := A e^{\delta T}$ we get

$$\begin{aligned}F_{LO}(0, s; \Phi) &= e^{-rT} \mathrm{E}_Q^{0,s}\Big(\Phi(S(T)) \cdot 1_{\{\inf S(t) e^{-\delta t} > A\}}\Big) \\ &= e^{-rT} \mathrm{E}_Q^{0,s}\Big(\Phi(\tilde{s}(T)) \cdot 1_{\{\inf \tilde{s}(t) > \tilde{L}\}}\Big).\end{aligned}$$

Note that $\tilde{L} = \mathit{const}$ and

$$d\tilde{s}(t) = \tilde{s}(t)\Big[(r - \delta)dt + \sigma dW(t)\Big], \qquad \tilde{s}_0 = \tilde{s}(0) = s_0 e^{\delta T}.$$

Under the measure Q the discounted price process $S(t)/e^{rt}$ is a martingale. Clearly, with the definition $\hat{s} := e^{\delta T} \cdot S$ the process $\hat{s}(t)/e^{rt}$ is also a Q-martingale. This implies the Q-martingale property of $\frac{\hat{s}(t)e^{-\delta t}}{e^{rt} \cdot e^{-\delta t}}$. Hence, in a fictitious market with interest rate $b := r - \delta$ the process $\tilde{s}(t)/e^{bt}$ is a Q-martingale so that we can apply Björk's arguments to get

$$e^{-bT} \mathrm{E}_Q^{0,s}\Big(\Phi(\tilde{s}(T)) \cdot 1_{\{\inf \tilde{s}(t) > \tilde{L}\}}\Big) = F_b(0, \tilde{s}_0; \Phi_{\tilde{L}}) - \left(\frac{\tilde{L}}{\tilde{s}_0}\right)^{\frac{2b}{\sigma^2} - 1} F_b\left(0, \frac{\tilde{L}^2}{\tilde{s}_0}; \Phi_{\tilde{L}}\right),$$

where the index b underlines that the prices have to be calculated with respect to the interest rate b. This leads to

$$\begin{aligned}F_{LO}(0, s; \Phi) &= e^{-\delta T} e^{-bT} \mathrm{E}_Q^{0,s}\Big(\Phi(\tilde{s}(T)) \cdot 1_{\{\inf \tilde{s}(t) > \tilde{L}\}}\Big) \\ &= e^{-\delta T} \cdot \left[F_b(0, \tilde{s}_0; \Phi_{\tilde{L}}) - \left(\frac{\tilde{L}}{\tilde{s}_0}\right)^{\frac{2b}{\sigma^2} - 1} F_b\left(0, \frac{\tilde{L}^2}{\tilde{s}_0}; \Phi_{\tilde{L}}\right)\right] \\ &= F(0, \tilde{s}_0; b, \Phi_{\tilde{L}}) - \left(\frac{\tilde{L}}{\tilde{s}_0}\right)^{\frac{2b}{\sigma^2} - 1} F\left(0, \frac{\tilde{L}^2}{\tilde{s}_0}; b, \Phi_{\tilde{L}}\right).\end{aligned}$$

The last equation results from the property that the price of a European derivative can be computed by discounting the Q-expectation of its final payment, i.e. $F_b(0, s; \Phi_L) = e^{-bT} \cdot \mathrm{E}_Q^{0,s}(\mathcal{Z}_{LO})$. □

An analogous result is true for up-and-out barrier derivatives.

Proposition 4.4 (Pricing Up-and-out Contracts) *Assume an exponential barrier (4.1). If we consider a T-claim $\mathcal{Z} = \Phi(S(T))$, then the pricing function of the corresponding up-and-out contract $\mathcal{Z}^{LO}$ is given, for $s < L$, by*

$$F^{LO}(t, s; \Phi) = F(t, \tilde{s}; b, \Phi^{\tilde{L}}) - \left(\frac{\tilde{L}}{\tilde{s}}\right)^{\frac{2b}{\sigma^2}-1} F\left(t, \frac{\tilde{L}^2}{\tilde{s}}; b, \Phi^{\tilde{L}}\right),$$

where $\tilde{s} := s \cdot e^{\delta(T-t)}$, $\tilde{L} := A \cdot e^{\delta T}$, and $b := r - \delta$. The price given by the pricing function $F(\cdot, \cdot; b, \cdot)$ has to be calculated as if the drift of the stock were equal to b.

Proof. Similar arguments apply as in the proof of Proposition 4.3. □

Example 4.1: "Case $\delta = r$"
In the special case $\delta = r$ equation (4.2) reads

$$F_{LO}(t, s; \Phi) = F(t, \tilde{s}; b, \Phi_{\tilde{L}}) - \frac{\tilde{s}}{\tilde{L}} F\left(t, \frac{\tilde{L}^2}{\tilde{s}}; b, \Phi_{\tilde{L}}\right),$$

where $\tilde{s} = s \cdot e^{r(T-t)}$, $\tilde{L} = A \cdot e^{rT}$, and $b = 0$. Hence, in the case of a down-and-out call the pricing function $F(\cdot, \cdot; 0, \cdot)$ is equivalent to the futures formula derived by Black (1976). In the following section we discuss this point in the more general framework of a Gaussian interest rate model.

4.4 Discounted Barrier and Gaussian Interest Rates

By a Gaussian interest rate model we mean any model of the term structure in which all bond price volatilities follow deterministic functions. Examples are the extended Vasicek model of Hull/White (1990) or the two-factor models considered by Heath/Jarrow/Morton (1992). For ease of exposition, we will concentrate on one-factor models. Hence, under the risk-neutral measure Q we get the following price dynamics of a (default-free) bond with maturity T:

$$dP(t,T) = P(t,T)\Big[r(t)dt + \sigma_B(t)dW_1(t)\Big],$$

where σ_B is a deterministic right-continuous function of time and W_1 is a one-dimensional Brownian motion. The short rate process r is assumed to be adapted to the Brownian filtration generated by W_1. Additionally, r (σ_B) shall be integrable (square-integrable) with respect to time t.

We now consider a stock with price dynamics under Q given by

$$dS(t) = S(t)\Big[r(t)dt + \sigma_{S1}(t)dW_1(t) + \sigma_{S2}(t)dW_2(t)\Big],$$

where σ_{S1} and σ_{S2} are deterministic functions of time which are square-integrable with respect to time t. Moreover, σ_{S2} shall be bounded away from zero and (W_1, W_2) is a two-dimensional Brownian motion. Assuming the *discounted boundary*

$$L(t) := A \cdot P(t,T), \quad A \in I\!R, \tag{4.3}$$

we obtain the following proposition:

Proposition 4.5 (Pricing Down-and-out Contracts) *If we consider a* T*-claim* $\mathcal{Z} = \Phi(S(T))$*, then the pricing function of the corresponding down-and-out contract* $\mathcal{Z}_{LO}$ *is given, for* $s > L(t)$*, by*

$$\begin{aligned} F_{LO}(t,s;T,\Phi) = P(t,T) \cdot \Bigg[& F_{0,1}\left(t, \frac{s}{P(t,T)}; \beta(t,T), \Phi_A\right) \\ & - \frac{s}{L(t)} F_{0,1}\left(t, \frac{L^2(t)}{s \cdot P(t,T)}; \beta(t,T), \Phi_A\right)\Bigg], \end{aligned}$$

where the notation $F_{0,1}(t,s;\beta(t,T),\Phi_A)$ *means that the price has to be calculated as if the interest rate were zero and as if the stock price* S *depended only on one Brownian motion with volatility equal to one. Besides, the time to maturity* $T-t$ *has to be replaced by*

$$\beta(t,T) := \int_t^T \Big[(\sigma_{S1}(s) - \sigma_B(s))^2 + \sigma_{S2}^2(s)\Big]\, ds.$$

As a result, one can calculate the price of a barrier derivative as in the futures model by Black (1976) with volatility equal to one and maturity equal to $\beta(t,T)$. Note that our proposition easily generalizes to multi-dimensional Gaussian interest rate models.

Proof. Without loss of generality we may set $t = 0$. Assume that $s = s_0 = S(0) > L(0)$. Again, to shorten notations, we write inf instead of $\inf_{0\leq t\leq T}$. Besides, $M(t)$ denotes the value of the money market account at time t, i.e. $M(t) = \exp(\int_0^t r(s)\,ds)$. We emphasize that $P(T,T) = 1$. Then, using the risk neutral valuation principle and performing a change to the T-forward measure Q^T, we arrive at[1]

$$\begin{aligned} F_{LO}(0,s;T,\Phi) = \mathrm{E}_Q^{0,s}\left(\frac{\mathcal{Z}_{LO}}{M(T)}\right) &= \mathrm{E}_Q^{0,s}\left(\frac{\Phi(S(T))}{M(T)} \cdot 1_{\{S(t)>L(t)\ \forall\ 0\leq t\leq T\}}\right) \\ &= P(0,T)\cdot \mathrm{E}_{Q^T}^{0,s}\left(\frac{\Phi(S(T))}{P(T,T)} \cdot 1_{\{S(t)>L(t)\ \forall\ 0\leq t\leq T\}}\right) \\ &= P(0,T)\cdot \mathrm{E}_{Q^T}^{0,s}\left(\Phi\left(\frac{S(T)}{P(T,T)}\right) \cdot 1_{\{\inf S(t)/P(t,T)>A\}}\right) \end{aligned}$$

with

$$d\left(\frac{S(t)}{P(t,T)}\right) = \frac{S(t)}{P(t,T)}\left[(\sigma_{S1}(t) - \sigma_B(t))dW_1^T(t) + \sigma_{S2}(t)dW_2^T(t)\right]$$

or equivalently,

$$\frac{S(t)}{P(t,T)} = \frac{s}{P(0,T)} \cdot \exp\Bigg(-0.5\int_0^t (\sigma_{S1}(s) - \sigma_B(s))^2 + \sigma_{S2}^2(s)\,ds \\ \underbrace{\int_0^t (\sigma_{S1}(s) - \sigma_B(s))\,dW_1^T(s) + \int_0^t \sigma_{S2}(s)\,dW_2^T(s)}_{=:X(t)}\Bigg).$$

Here, the process (W_1^T, W_2^T) is a Brownian motion with respect to the T-forward measure Q^T. Let

$$\beta(t) := \beta(0,t) :=< X >_t = \int_0^t \left[(\sigma_{S1}(s) - \sigma_B(s))^2 + \sigma_{S2}^2(s)\right] ds.$$

Since σ_{S2} is bounded away from zero, the function β is strictly increasing, which implies that the stopping time $\tau(s) := \inf\{t \geq 0 : \beta(t) > s\}$ is the left- and right-inverse of β. Using the theorem of time-change for martingales, we get[2]

$$\frac{S(\tau(t))}{P(\tau(t),\tau(T))} = \frac{s}{P(0,\tau(T))} \cdot \exp\Big(-0.5t + \tilde{W}^T(t)\Big), \quad 0 \leq t \leq T,$$

[1] See Geman/El Karoui/Rouchet (1995).
[2] See Karatzas/Shreve (1991), p. 174.

where $\tilde{W^T}$ denotes a one-dimensional Brownian motion with respect to Q^T. Hence, we can apply Björk's result with $r = 0$ and $\sigma = 1$, which leads to

$$\mathrm{E}_{Q^T}^{0,s}\left(\Phi\left(\frac{S(\tau(T))}{P(\tau(T),\tau(T))}\right)\cdot 1_{\{\inf\limits_{0\leq t\leq\tau(T)} S(t)/P(t,\tau(T))>A\}}\right)$$
$$= F_{0,1}\Big(0, s/P(0,\tau(T)); T, \Phi_A\Big) - \tfrac{s}{A\cdot P(0,\tau(T))}F_{0,1}\Big(0, A^2P(0,\tau(T))/s; T, \Phi_A\Big).$$

Since $\tau(\beta(T)) = T$, we arrive at

$$\mathrm{E}_{Q^T}^{0,s}\left(\Phi\left(\frac{S(T)}{P(T,T)}\right)\cdot 1_{\{\inf\limits_{0\leq t\leq T} S(t)/P(t,T)>A\}}\right)$$
$$= F_{0,1}\Big(0, s/P(0,T); \beta(T), \Phi_A\Big) - \tfrac{s}{A\cdot P(0,T)}F_{0,1}\Big(0, A^2\cdot P(0,T)/s; \beta(T), \Phi_A\Big).$$

This completes the proof. □

Example 4.1 : "Down-and-out Call"

Consider the case of a down-and-out call with contract function $\Phi(x) = \max\{x - K; 0\}$, $K \geq A$, and choose $s > A$. Applying Proposition 4.5 we get

$$\begin{aligned}
C_{LO}(t,s;T) &:= F_{LO}(t,s;T,\Phi)\\
&= P(t,T)\Big[F_{0,1}\Big(t, s/P(t,T); \beta(t,T), \Phi_A\Big)\\
&\qquad -\frac{s}{L(t)}F_{0,1}\Big(t, L^2(t)/(s\cdot P(t,T)); \beta(t,T), \Phi_A\Big)\Big]\\
&= C(t,s;K) - \frac{s}{L(t)}\cdot C(t, L^2(t)/s; K)\\
&= C(t,s;K) - \frac{K}{A}\cdot Put(t,s;A^2/K)
\end{aligned} \tag{4.4}$$

where

$$C(t,s;K) = s\cdot N(d_1(t,s;K)) - K\cdot P(t,T)\cdot N(d_2(t,s;K)),$$
$$Put(t,s;K) = K\cdot P(t,T)\cdot N(-d_2(t,s;K)) - s\cdot N(-d_1(t,s;K))$$

with

$$d_1(t,s;K) = \frac{\ln(s/K) - \ln P(t,T) + 0.5\beta(t,T)}{\sqrt{\beta(t,T)}},$$
$$d_2(t,s;K) = d_1(t,s) - \sqrt{\beta(t,T)}.$$

As proved by Merton (1973) and Amin/Jarrow (1992), the function $C(t,s;K)$ corresponds to the price of a European call in a generalized version of the

Black-Scholes model with a Gaussian term structure. From (4.4) it becomes obvious that the down-and-out call can be statically hedged in terms of Bowie/Carr (1994) using one ordinary call and K/A ordinary puts with strike A^2/K. For $K = A$ we get $C_{LO}(0, s_0, T) = s_0 - K \cdot P(0,T)$. This latter result does not depend on any specific model of the stock prices and can also be proved using a no-arbitrage argument.

In this "down-and-out world" with discounted barrier the prices of other barrier derivatives also have interesting representations. We will show this in the following examples.

Example 4.2: "Down-and-out Bond"

A down-and-out bond expires valueless if a lower barrier L is touched during its lifetime, otherwise the payoff of the down-and-out bond equals the payoff of an ordinary bond. In this example we calculate the price of a down-and-out bond. For the sake of simplicity, let $t = 0$. In the case of interest rates equal to zero and volatility equal to one, the price of a binary call with strike A is given by

$$H_{0,1}(0, s_0; T, A) = N\left(\frac{\ln(\frac{s_0}{A}) - 0.5T}{\sqrt{T}}\right).$$

Let $BO \equiv 1$ denote the contract function of a bond. Applying Proposition 4.5 leads to

$$\begin{aligned}
P_{LO}(0, s_0; T) &:= F_{LO}(0, s_0, ; T, BO) \\
&= p_0 \cdot \left[F_{0,1}\left(0, \tfrac{s_0}{p_0}; \beta_0, BO_A\right) - \frac{s_0}{Ap_0} \cdot F_{0,1}\left(0, \tfrac{L^2(0)}{s_0 \cdot p_0}; \beta_0, BO_A\right)\right] \\
&= p_0 \cdot \left[H_{0,1}\left(0, \tfrac{s_0}{p_0}; \beta_0, A\right) - \frac{s_0}{Ap_0} \cdot H_{0,1}\left(0, \tfrac{L^2(0)}{s_0 \cdot p_0}; \beta_0, A\right)\right] \\
&= p_0 \cdot \left[N\left(\frac{\ln(\frac{s_0}{Ap_0}) - 0.5\beta_0}{\sqrt{\beta_0}}\right) - \frac{s_0}{Ap_0} \cdot N\left(\frac{\ln(\frac{Ap_0}{s_0}) - 0.5\beta_0}{\sqrt{\beta_0}}\right)\right] \\
&= \frac{1}{A} \cdot \left[A \cdot p_0 \cdot N(\hat{d}_2) - s_0 \cdot N(-\hat{d}_1)\right]
\end{aligned}$$

with

$$\hat{d}_1 = \frac{\ln(\frac{s_0}{Ap_0}) + 0.5\beta_0}{\sqrt{\beta_0}}, \qquad \hat{d}_2 = \hat{d}_1 - \sqrt{\beta_0}.$$

Hence, the price of a down-and-out bond is the same as the price of $1/A$ ordinary puts except for the sign of the variable $\hat{d}_2$.

Example 4.3: "Down-and-out Stock"
Using the results of both the above examples, one can calculate the price of a down-and-out stock. Let $ST(x) = x$ be the contract function of a stock, then we obtain

$$ST_A(x) = A \cdot CB(x; A) + CA(x; A), \qquad x \in I\!R,$$

where $CB(\cdot; A)$ and $CA(\cdot; A)$ denote the contract functions of a binary call and an ordinary call with strike A, respectively. Since a pricing function is linear in the contract function, we can calculate the pricing function of a down-and-out stock:

$$\begin{aligned} S_{LO}(0, s_0; T) &:= F_{LO}(0, s_0; T, ST) \\ &= A \cdot P_{LO}(0, s_0; T) + C_{LO}(0, s_0; T) \\ &= A \cdot p_0 \cdot N(\hat{d}_2) - s_0 \cdot N(-\hat{d}_1) + s_0 - A \cdot p_0 \\ &= s_0 \cdot N(\hat{d}_1) - A \cdot p_0 \cdot N(-\hat{d}_2). \end{aligned}$$

Consequently, the price of a down-and-out stock is similar to the price of an ordinary call except for the sign of the variable $\hat{d}_2$. This completes the example.

Proposition 4.5 has an analogue for up-and-out contracts:

Proposition 4.6 (Pricing Up-and-out Contracts) *If we consider a T-claim $\mathcal{Z} = \Phi(S(T))$, then the pricing function of the corresponding up-and-out contract $\mathcal{Z}^{LO}$ is given, for $s < L(t)$, by*

$$\begin{aligned} F^{LO}(t, s; T, \Phi) = P(t,T) \cdot \Bigg[& F_{0,1}\left(t, \frac{s}{P(t,T)}; \beta(t,T), \Phi^A\right) \\ & - \frac{s}{L(t)} F_{0,1}\left(t, \frac{L^2(t)}{s \cdot P(t,T)}; \beta(t,T), \Phi^A\right)\Bigg], \end{aligned}$$

where the notation $F_{0,1}(t, s; \beta(t,T), \Phi^A)$ means that the price has to be calculated as if the interest rate were zero and as if the stock price S depended only on one Brownian motion with volatility equal to one. Besides, the time to maturity $T - t$ has to be replaced by

$$\beta(t,T) := \int_t^T \left[(\sigma_{S1}(s) - \sigma_B(s))^2 + \sigma_{S2}^2(s)\right] ds.$$

Using our above results, one can calculate the prices of an up-and-out bond and an up-and-out stock.

Example 4.4: "Up-and-out Bond and Up-and-out Stock"
Given that $s_0 < A \cdot p_0$ we arrive at:

$$\begin{aligned} P^{LO}(0, s_0; T) &:= F^{LO}(0, s_0; T, BO) \\ &= \frac{1}{A} \cdot \Big[A \cdot p_0 \cdot N(-\hat{d}_2) - s_0 \cdot N(\hat{d}_1) \Big] \end{aligned}$$

$$\begin{aligned} S^{LO}(0, s_0; T) &:= F^{LO}(0, s_0; T, ST) \\ &= A \cdot P^{LO}(0, s_0, T) - Put^{LO}(0, s_0, T) \\ &= A \cdot p_0 \cdot N(-\hat{d}_2) - s_0 \cdot N(\hat{d}_1) - A \cdot p_0 + s_0 \\ &= s_0 \cdot N(-\hat{d}_1) - A \cdot p_0 \cdot N(\hat{d}_2). \end{aligned}$$

Since $N(-\hat{d}_2) > N(\hat{d}_1)$ both results are positive.

4.5 Application: Pricing of Defaultable Bonds

Prominent examples where derivatives with curved boundaries play an important role are several firm value models for the pricing of corporate liabilities. In Chapter 5 we will discuss these models in more detail. Here, we only consider the model of Briys/de Varenne (1997) and show how Proposition 4.5 can be applied to find the value of corporate debt in their model. It is worth mentioning that Proposition 4.3 is tailored to determine the value of debt in the model of Black/Cox (1976).

In the Briys-de Varenne model the value of the firm follows the dynamics

$$dV(t) = V(t) \Big[r(t)dt + \sigma_V(t)dW_V(t) + \sigma_r(t)dW_r(t) \Big], \quad V(0) = v_0,$$

under a martingale measure Q, where (W_V, W_r) is a two-dimensional Brownian motion under this measure. The processes $\sigma_V(t) = \sigma_V(t, V(t), r(t))$ and $\sigma_r(t) = \sigma_r(t, V(t), r(t))$ are supposed to be measurable and sufficiently integrable so that the SDE has a unique solution. Briys/de Varenne (1997) analyzed a model which is complete and free of arbitrage. Therefore, the measure Q is unique. The short rate r is modeled using an extended Vasicek model, i.e. under the martingale measure Q the short rate dynamics are governed by

$$dr(t) = a(t)dt + bdW_r(t), \quad r(0) = 0, \tag{4.5}$$

with $a(t) = \theta(t) - \alpha r(t)$, $\alpha > 0$, and $b > 0$. In contrast to Vasicek's model, where θ is a constant, a deterministic function θ is introduced, which can be used to calibrate the model to the initial term structure.

Applying the approach of Merton (1974), Briys/de Varenne (1997) modeled corporate debt as contingent claim on the value of the firm and assume that debt consists only of a single zero bond. Three scenarios can occur in their model: If the firm does well during the lifetime of the bond and the value of the firm at the bond's maturity T_B is greater than the face value F of the bond, the bondholders receive the face value at T_B. In this context, "doing well" means that during the lifetime of the bond the value of the firm does not touch a lower barrier

$$L(t) = k \cdot F \cdot P(t, T_B),$$

where $0 \leq k \leq 1$ and $P(t, T_B)$ denotes the price of a riskless zero bond with maturity T_B at time t. If the firm does well, but at maturity the value of the firm is smaller than the face value of debt, the bondholders receive the assets of the firm. Given a premature default, that means the firm value drops to the barrier before maturity, the bondholders also receive the firm's. Therefore, the value of debt at its maturity T_B is given by[3]

$$B(T_B, V(T_B)) = F \cdot P_{LO}(T_B) - \max\{F - V(T_B), 0\}_{LO} + k \cdot F \cdot P_{LI}(T_B),$$

where P_{LI} denotes a down-and-in bond with knock-in barrier L and P_{LO} denotes the corresponding down-and-out bond. Therefore, we have $P_{LO}(T_B) = 1_{\{\tau > T_B\}}$ and $P_{LI}(T_B) = 1_{\{\tau \leq T_B\}}$, where 1_E stands for the indicator function for the event E and $\tau = \inf\{t \geq 0 : V(t) = L(t)\}$. Besides, $\max\{F - V(T_B), 0\}_{LO}$ denotes a down-and-out put with strike F and knock-out barrier L.

To apply our results of Section 4.4, let us rewrite the payoff profile using the in-out parity and the put-call parity for down-and-out options. We thus obtain

$$B(T_B, V(T_B)) = S_{LO}(T_B) - C_{LO}(T_B) + k \cdot F \cdot (1 - P_{LO}(T_B)).$$

[3] Briys/de Varenne (1997) also analyze deviations from the absolute priority rule. We shall not consider this generalization here because it does not fundamentally alter our results.

Hence, we need to compute the price of a down-and-out stock S_{LO}, a down-and-out call C_{LO} with strike F, and a down-and-out bond. Let $p_0 = P(0, T_B)$ and $\beta_0 = \beta(0, T_B)$. According to our Examples 4.1, 4.2, and 4.3 it follows

$$\begin{aligned}
S_{LO}(0) &= v_0 \cdot N(\hat{d}_1) - k \cdot F \cdot p_0 \cdot N(-\hat{d}_2), \\
C_{LO}(0) &= v_0 \cdot N(d_1) - F \cdot p_0 \cdot N(d_2) + F \cdot p_0 \cdot N(-\tilde{d}_2) - v_0 \cdot N(-\tilde{d}_1), \\
P_{LO}(0) &= p_0 \cdot N(\hat{d}_2) - \tfrac{v_0}{kF} \cdot N(-\hat{d}_1)
\end{aligned}$$

with

$$\begin{aligned}
d_1 &= \frac{\ln(\frac{v_0}{Fp_0}) + 0.5\beta_0}{\sqrt{\beta_0}}, \quad d_2 = d_1 - \sqrt{\beta_0}, \\
\tilde{d}_1 &= \frac{\ln(\frac{v_0}{k^2 F p_0}) + 0.5\beta_0}{\sqrt{\beta_0}}, \quad \tilde{d}_2 = \tilde{d}_1 - \sqrt{\beta_0}, \\
\hat{d}_1 &= \frac{\ln(\frac{s_0}{kFp_0}) + 0.5\beta_0}{\sqrt{\beta_0}}, \quad \hat{d}_2 = \hat{d}_1 - \sqrt{\beta_0}.
\end{aligned}$$

Therefore, we obtain the following value of debt at time $t = 0$

$$B(0) = F \cdot p_0 \Big[1 - N(-d_2) + \tfrac{v_0}{Fp_0} \cdot N(-d_1) + k \cdot N(-\tilde{d}_2) - \tfrac{v_0}{kFp_0} N(-\tilde{d}_1) \Big].$$

Since the price of the riskless bond equals $F \cdot p_0$, we can interpret the last four terms in the brackets as correction terms. These terms comes from the fact that, in the event of default, the debtholders receive only portions of the promised payments.

4.6 Conclusion

In this chapter we have proved valuation formulae for barrier derivatives with curved boundaries. These are generalizations of results by Björk (1998). In case of deterministic interest rates we have considered a deterministic exponential boundary. Given a Gaussian interest rate model, a discounted boundary has been assumed and closed-form solutions for the prices of barrier derivatives have been computed. We emphasize that in more general settings, as e.g. considered by Roberts/Shortland (1997), we cannot hope to find closed-form solutions for the prices of barrier derivatives. This is so because, in general, there does not exist a closed-form solution for the transition density of the price process of the underlying.

Moreover, we have discussed relevant examples and applied our results to obtain the value of corporate debt in the firm value framework of Briys/de

Varenne (1997). Clearly, this value is well-known from literature. Nevertheless, our approach to decompose the bond into derivatives with discounted boundaries and then to price these building blocks is a generalization of ideas presented by Ericsson/Reneby (1998). Whereas these authors assumed deterministic interest rates, we have considered the framework of Briys/de Varenne (1997).

5

Optimal Portfolios with Defaultable Assets - A Firm Value Approach

5.1 Introduction

In his pioneering work Merton (1969, 1971) considered an investor who allocates wealth to stocks and to a riskless money market account. However, the assumption is made that the interest rates are deterministic and that all assets are free of default risk. Relaxing the first point has already been addressed in Chapter 2. The second point is rarely treated in literature.[1]

In this chapter we will solve portfolio problems where the investor can put funds into defaultable assets such as corporate bonds. To model default risk we use a firm value approach, which goes back to Black/Scholes (1973) and Merton (1974). They developed the so-called Merton's (firm value) model in which firm value serves as main ingredient. By definition, firm value equals the sum of the firm's liabilities. If, for example, a firm has not issued any debt, firm value coincides with the value of equity. Merton considered a firm that also issued debt in the form of a single zero bond. In the event that firm value at the maturity of the zero bond is larger than the face value of debt, the stockholders redeem the bond. Otherwise, the bond defaults and the firm's assets pass over to the bondholders. Therefore, the zero bond can be interpreted as a portfolio consisting of an amount in cash payable at the bond's maturity and a short position in a put option on firm value. The exercise price of this put is equal to the face value of the bond. Equity corresponds to a call

[1] To our knowledge, only Merton (1971) considered a portfolio problem with defaultable bonds, but he used a bond model which can be seen as a rudimentary reduced form model with deterministic interest rates.

on firm value, struck at the face value of the bond. In this framework default can only occur at the maturity of the bond. In reality, bond indentures often contain special provisions such as safety covenants, which give the bondholder the right to declare bankruptcy if the firm is doing poorly according to some standard. In firm value models this is modeled by a bankruptcy level which can be time dependent or even stochastic. The firm is forced to bankruptcy when firm value falls to the bankruptcy level. This is worked out by Black/Cox (1976) for deterministic interest rates and by Briys/de Varenne (1997) for stochastic interest rates. Black/Cox (1976) also considered the case of subordinated bonds and restrictions on the financing of interest and dividend payments. Multiple further generalizations were addressed in Geske (1977), Mason/Bhattacharya (1981), Kim/Ramaswamy/Sundaresan (1993), Leland (1994), Leland/Toft (1996), Longstaff/Schwartz (1995), Saa-Requejo/Santa-Clara (1999), Collin-Dufresne/Goldstein (2001). A recent empirical investigation of some of these models was worked out by Eom/Helwege/Huang (2002).

Let us now come to the portfolio problem which we are going to address in this chapter. We assume that the investor wishes to put his wealth into shares of the firm's equity (syn. stocks), into corporate bonds, and into a riskless money market account.[2] Actually, this means that the investor has the choice between two derivatives and the money market account. The main restriction occurring in this problem results from the fact that the capital structure of the firm is explicitly modeled because firm value equals the sum of equity and debt. If, for the sake of simplicity, the total number of stocks is normalized to unity, the investor can buy at most one share of equity and one share of debt, namely the single zero bond. In the ordinary formulation of portfolio problems such restrictions are not included because capital structures are not modeled in an explicit way.

The chapter is organized as follows. In Section 5.2 we will first take up the simplifying approach of the usual problems and ignore the upper bounds on the numbers of stocks and bonds. This can be justified as a first order approximation for a small investor whose total wealth will (almost) never be sufficient to buy the entire issues of equity or debt. After having solved this problem, we will look in Section 5.3 at the probability that the solution com-

[2] Strictly speaking, investing in bonds means investing in shares of the firm's debt which, by assumption, consists of a single bond.

puted without the above constraints violates these constraints. In Section 5.4 we solve the general constrained problem under the assumption that the investor maximizes utility from terminal wealth with respect to a logarithmic utility function. The chapter ends with a conclusion.

5.2 The Unconstrained Case

As mentioned above, it is a fact inherent in firm value models that the total numbers of stocks and bonds are limited to one. Using the "small investor assumption" as a justification, we will first ignore this fact and assume that there are no upper bounds on the total numbers of stocks and bonds. We will consider Merton's model and the Black-Cox model.

5.2.1 Merton Model

Since a defaultable bond can be interpreted as a portfolio consisting of an amount in cash F payable at the bond maturity T_B and a short position in a put with exercise price F, its value at T_B is given by

$$B(T_B, v) = \min\{v, F\} = F - \max\{F - v, 0\}, \qquad v \geq 0.$$

The value of the stock equals the value of a call on firm value with exercise price F and maturity T_B, i.e.

$$S(T_B, v) = \max\{v - F, 0\}, \qquad v \geq 0.$$

Throughout the chapter we make the assumption that stocks and bonds are continuously traded on a frictionless market. This does not mean that the investor can actually trade in both assets, but it simplifies our presentations because, under this assumption, Black-Scholes-like formulae for the stock and the bond price are valid. If one or both of these assets were not traded, a market price of risk would come into play. Although this would be an unpleasant feature for contingent claim pricing, it is a usual situation in the context of portfolio optimization. For this reason, our assumption can be made without loss of generality.

Optimizing a portfolio consisting of defaultable bonds can be considerably simplified if we apply the *elasticity approach to portfolio optimization*, which has been developed in Chapter 3. Basically, this approach says that the optimal wealth process is determined by an optimal elasticity which is independent of a specific asset. Therefore, it consists of a kind of *two-step procedure*: First determine the optimal elasticity which characterizes the optimal investor's wealth process (for a given utility function) and then compute the portfolio process that leads to this elasticity (for given tradable assets).

Hence, we start with a portfolio problem where the investor can put his wealth into the money market account and the value of the firm modeled by the SDEs

$$dM(t) = M(t)rdt, \qquad M(0) = 1, \tag{5.1}$$

$$dV(t) = V(t)\Big[\alpha dt + \sigma dW(t)\Big], \qquad V(0) = v_0. \tag{5.2}$$

The variable M denotes the balance of the money market account and V the value of the firm which has a constant drift α and a constant volatility σ. Here and in the following W stands for a Brownian motion defined on a filtered probability space $(\Omega, \mathcal{F}, P)$. The filtration $\{\mathcal{F}_t\}_{t\geq 0}$ is the usual P-augmentation of the natural filtration of W. In this section W is one dimensional. The variable r stands for the short rate which is held constant to simplify matters. Let $\lambda := \alpha - r$ be the excess return on the firm-value process.

Actually, the value of the firm is not tradable. However, the elasticity approach tells us that we can solve the portfolio problem with the two investment opportunities V and M, and then compute how the optimal wealth process can be attained by investing in tradable securities, i.e. in stocks and bonds issued by the firm. As a consequence, we look at the following wealth equation

$$dX(t) = X(t)\Big[(r + \pi_V(t)\lambda)dt + \pi_V(t)\sigma dW(t)\Big], \tag{5.3}$$

$X(0) = x_0$, where X denotes the investor's total wealth and π_V stands for the percentage of total wealth put into the value of the firm.

The classical examples of the portfolio problem

$$\max_{\pi} \mathrm{E}(U(X^{\pi}(T))) \tag{5.4}$$

are the choices of $U(x) = \ln(x)$ or $U(x) = \frac{1}{\gamma}x^{\gamma}$. The variable T denotes the investment horizon. We assume that $T < T_B$ which means that during the investment period $[0, T]$ a default cannot occur. Of course, a low value of the firm indicates a high probability of default and a low value of debt. In the following subsections this assumption will be relaxed.

The optimal portfolio processes for (5.4) are well-known from the seminal papers by Merton (1969, 1971).

Proposition 5.1 (Merton's Portfolio Problem)
(i) For a logarithmic utility function $U(x) = \ln(x)$ the optimal portfolio process π_V^* for (5.4) is given by

$$\pi_V^*(t) = \frac{\lambda}{\sigma^2}.$$

(ii) For a power utility function $U(x) = \frac{1}{\gamma}x^\gamma$, $\gamma < 1$, $\gamma \neq 0$, the optimal portfolio process π_V^* for (5.4) is given by

$$\pi_V^*(t) = \frac{\lambda}{(1-\gamma)\cdot\sigma^2}.$$

However, the value of the firm V shall not be tradable, whereas claims on V are. Although their prices are both non-linear functions of firm value, we can still use the results of the above proposition to obtain the optimal wealth process. To demonstrate this idea, assume that the investor can additionally invest money in a contingent claim $C(t) = C(t, V(t))$ on firm value. An application of Ito's rule and of the Black-Scholes partial differential equation results in the SDE

$$dC = (rC + C_v V\lambda)dt + C_v V\sigma dW$$

for the claim price. Here, C_v denotes the partial derivative of $C(t, v)$ with respect to firm value. The wealth equation of this portfolio problem is given by

$$dX = X\Big[(r + (\pi_V + \pi_C C^{-1}C_v V)\lambda)dt + (\pi_V + \pi_C C^{-1}C_v V)\sigma dW\Big],$$

where π_V denotes the percentage invested in the claim. This equation involves the elasticity of the claim with respect to firm value which is defined by

$$\varepsilon_C = \frac{dC/C}{dV/V} := \frac{C_v V}{C}.$$

Note that for the corresponding elasticities of firm value and of the money market account we have $\varepsilon_V \equiv 1$ and $\varepsilon_M \equiv 0$, respectively. Therefore, the term

$$\varepsilon := \pi_V + \pi_C C^{-1}C_v V = \pi_V\varepsilon_V + \pi_C\varepsilon_C$$

coincides with the static elasticity of the investor's portfolio.[3] Using this result, the wealth equation simplifies to

$$dX = X\Big[(r + \varepsilon\lambda)dt + \varepsilon\sigma dW\Big]. \tag{5.5}$$

In this formulation the static portfolio elasticity ε is the control variable of the portfolio problem. Note that ε does not depend on a specific asset. Since the wealth equations (5.3) and (5.5) only differ with respect to the notation of the control variable, investment problems with contingent claims of the form $C(t) = C(t, V(t))$ can be solved as if the value of the firm were tradable and as if the investor could only invest in the value of the firm and the money market account. This simpler problem is said to be the *reduced portfolio problem*. For example, if the investor maximizes utility from terminal wealth at time T with respect to a logarithmic utility function $U(x) = \ln(x)$, $x > 0$, the above proposition yields that the optimal elasticity reads

$$\varepsilon^*(t) = \frac{1}{1-\gamma}\frac{\lambda}{\sigma^2}.$$

Hence, any combination of firm value and claim which leads to the optimal elasticity ε^* can be selected to achieve the optimal wealth process. This is the main result of the second step of the elasticity approach. Formally, we find that (π_V, π_C) has to be chosen so that

$$\varepsilon^*(t) = \pi_V(t) + \pi_C(t) \cdot \varepsilon_C(t).$$

As in our case the value of the firm V is not tradable, we must have $\pi_V \equiv 0$, which leads to

$$\varepsilon^*(t) = \pi_C(t) \cdot \varepsilon_C(t)$$

or

$$\pi_C^*(t) = \frac{\varepsilon^*(t)}{\varepsilon_C(t)}. \tag{5.6}$$

Since stocks and bonds in Merton's model are contingent claims on firm value, we obtain the following proposition:

Proposition 5.2 *Consider the portfolio problem (5.4) and Merton's firm value model.*

[3] The word "static" emphasizes that ε only equals the elasticity of the portfolio if π is held constant. Otherwise, Ito's rule needs to be applied and additional terms come into play.

(i) If the investor can only invest in the money market account and stocks issued by the firm, then the optimal stock portfolio process π_S^* is given by

$$\pi_S^*(t) = \begin{cases} \frac{\lambda}{\sigma^2} \cdot \frac{S(t)}{S_v(t) \cdot V(t)} = \frac{\lambda}{\sigma^2} \cdot \frac{S(t)}{\mathcal{N}(d_1(t)) \cdot V(t)} & \text{if } U(x) = \ln(x), \\ \frac{\lambda}{(1-\gamma)\sigma^2} \cdot \frac{S(t)}{S_v(t) \cdot V(t)} = \frac{\lambda}{(1-\gamma)\sigma^2} \cdot \frac{S(t)}{\mathcal{N}(d_1(t)) \cdot V(t)} & \text{if } U(x) = \frac{1}{\gamma}x^\gamma, \end{cases}$$

where S denotes the stock price in Merton's model, i.e.

$$S(t) = V(t) \cdot \mathcal{N}(d_1(t)) - Fe^{-r(T-t)}\mathcal{N}(d_2(t))$$

with

$$d_1(t) = \frac{\ln(\frac{V(t)}{F}) + (r + 0.5\sigma^2)(T-t)}{\sigma\sqrt{T-t}},$$

$$d_2(t) = d_1(t) - \sigma\sqrt{T-t},$$

and $\mathcal{N}$ denotes the cumulative standard normal distribution function.

(ii) If the investor can only invest in the money market account and bonds issued by the firm, then the optimal bond portfolio process π_B^* is given by

$$\pi_B^*(t) = \begin{cases} \frac{\lambda}{\sigma^2} \cdot \frac{B(t)}{B_v(t) \cdot V(t)} = \frac{\lambda}{\sigma^2} \cdot \frac{B(t)}{\mathcal{N}(-d_1(t)) \cdot V(t)} & \text{if } U(x) = \ln(x), \\ \frac{\lambda}{(1-\gamma)\sigma^2} \cdot \frac{B(t)}{B_v(t) \cdot V(t)} = \frac{\lambda}{(1-\gamma)\sigma^2} \cdot \frac{B(t)}{\mathcal{N}(-d_1(t)) \cdot V(t)} & \text{if } U(x) = \frac{1}{\gamma}x^\gamma, \end{cases}$$

where B denotes the bond price in Merton's model, i.e.

$$B(t) = V(t) \cdot \mathcal{N}(-d_1(t)) + Fe^{-r(T-t)}\mathcal{N}(d_2(t))$$

with d_1 and d_2 defined as in (i).

(iii) If the investor can put his wealth into stocks, bonds, and the money market account, then every portfolio process (π_S, π_B) is optimal if it leads to optimal elasticity, i.e.

$$\varepsilon^* = \pi_S \cdot \varepsilon_S + \pi_B \cdot \varepsilon_B.$$

Hence, the optimal strategy is not uniquely determined.

Proof. The proof is a direct consequence of Proposition 5.1, the form of the optimal elasticity ε^* in (5.6) and the price formulae for both stock and bond issued by the firm. □

Remarks.

a) By applying the concept of elasticity to portfolio optimization, we generalize results of Korn/Trautmann (1999), who solved portfolio problems with options.

b) For expositional convenience we have only considered an investor with logarithmic or power utility function. Clearly, the results of Proposition 5.2 are neither restricted to these utility functions nor to the maximization of utility from terminal wealth, but they hold for every utility function considered in Merton (1969,1971), Cox/Huang (1989, 1991) or Karatzas/Lehoczky/Shreve (1987).

c) By comparing the actual amounts of money invested in firm value and debt as computed in Proposition 5.1 and 5.2, we find

$$\pi_V^* \cdot X^* = \frac{\lambda}{\sigma^2} X^* > \frac{\lambda}{\sigma^2} \frac{B}{B_v V} X^* = \pi_B^* \cdot X^*.$$

Consequently, the optimal amount of money invested in the risky asset will always be lower if we invest in debt than if we were able to invest in firm value. Therefore, there is less money under default risk, which seems to be a desirable feature. The same result holds for firm value and equity.

5.2.2 On the Assumption that Firm Value is Tradable

This subsection is intended as a short excursion on the assumption that firm value is tradable. It is often claimed that this assumption represents a fundamental drawback of firm value models. The aim of this subsection is to show that it can be replaced by the assumption that securities of the firm, such as stocks, are continuously traded.

As before, we assume that the firm has issued debt in form of a single zero bond B and that the shares of stock S are continuously traded. We emphasize that we do not assume that the investor is able to trade in the bond continuously, but the opportunity must exist to observe the bond price continuously and to buy or sell the bond at a certain point in time, e.g. if mispricing occurs.

The price processes follow the stochastic differential equations (SDE)

$$\begin{aligned} dS(t) &= S(t)\Big[\alpha_S(t)dt + \sigma_S(t)dW(t)\Big], \\ dB(t) &= B(t)\Big[\alpha_B(t)dt + \sigma_B(t)dW(t)\Big]. \end{aligned} \tag{5.7}$$

The value of the firm is given by definition, $V := S + B$. Further, we assume that there exist real-valued, continuous functions s and b such that $S(t) = s(t, V(t))$ and $B(t) = b(t, V(t))$. In addition, the values of the firm's liabilities at time T_B are given by

$$S(T_B) = \max\{V(T_B) - F; 0\},$$
$$B(T_B) = F - \max\{F - V(T_B); 0\},$$

where F denotes the face value of the bond. The firm-value process is represented by the following SDE:

$$dV(t) = dS(t) + dB(t) = V(t)\Big[\alpha(t)dt + \sigma(t)dW(t)\Big],$$

where

$$\alpha(t) = \frac{S(t)\alpha_S(t) + B(t)\alpha_B(t)}{V(t)}, \qquad \sigma(t) = \frac{S(t)\sigma_S(t) + B(t)\sigma_B(t)}{V(t)}.$$

We restrict our considerations to the case that $\sigma(t)$ is constant, which corresponds to the model analyzed by Merton (1974). Applying Ito's formula leads to alternative representations for the dynamics of S and B:

$$\begin{aligned} dS &= \Big(S_t + S_v V\alpha + 0.5 S_{vv} V^2\sigma^2\Big)dt + S_v V\sigma dW, \\ dB &= \Big(B_t + B_v V\alpha + 0.5 B_{vv} V^2\sigma^2\Big)dt + B_v V\sigma dW. \end{aligned} \tag{5.8}$$

If we consider a self-financing trading strategy $(\varphi_M, \varphi_S, \varphi_B)$ with $\varphi_B \equiv 1$, the wealth equation reads

$$\begin{aligned} dX^\varphi &= dB + \varphi_S dS + \varphi_M dM \\ &= \Big(B_t + B_v V\alpha + 0.5 B_{vv} V^2\sigma^2\Big)dt + B_v V\sigma dW \\ &\quad + \varphi_S\Big[\Big(S_t + S_v V\alpha + 0.5 S_{vv} V^2\sigma^2\Big)dt + S_v V\sigma dW\Big] + \varphi_M M r dt \\ &= \Big[B_t + B_v V\alpha + 0.5 B_{vv} V^2\sigma^2 \\ &\quad + \varphi_S\left(S_t + S_v V\alpha + 0.5 S_{vv} V^2\sigma^2\right) + \varphi_M M r\Big]dt + \Big[B_v + \varphi_S S_v\Big]\sigma V dW. \end{aligned}$$

Note that the strategy $\varphi_B \equiv 1$ does not require continuous trading in the bond. Choosing $\varphi_S = -B_v/S_v$ leads to a locally riskless position. Since the drift term is equal to

$$\begin{aligned} & B_t + B_v V\alpha + 0.5 B_{vv} V^2\sigma^2 - \tfrac{B_v}{S_v}\left(S_t + S_v V\alpha + 0.5 S_{vv} V^2\sigma^2\right) + \varphi_M M r \\ &= r\Big(\underbrace{\varphi_M M + B - \tfrac{B_v}{S_v} S}_{=X^\varphi}\Big) \\ &\quad - rB + r\tfrac{B_v}{S_v} S + B_t + 0.5 B_{vv} V^2\sigma^2 - S_v^{-1}\Big(B_v S_t + 0.5 B_v S_{vv} V^2\sigma^2\Big), \end{aligned}$$

it follows that

$$-rB + r\frac{B_v}{S_v}S + B_t + 0.5B_{vv}V^2\sigma^2 - S_v^{-1}\Big(B_vS_t + 0.5B_vS_{vv}V^2\sigma^2\Big) = 0,$$

or equivalently,

$$-rBS_v + rB_vS + B_tS_v + 0.5B_{vv}V^2\sigma^2S_v - B_vS_t - 0.5S_{vv}V^2\sigma^2B_v = 0. \tag{5.9}$$

The relation $V = B + S$ yields

$$\begin{aligned} S_v &= 1 - B_v, \qquad (5.10)\\ S_{vv} &= -B_{vv},\\ S_t &= -B_t. \end{aligned}$$

Substituting these conditions into equation (5.9) leads to the well-known Black-Scholes partial differential equation

$$B_t + rB_vV + 0.5B_{vv}V^2\sigma^2 - rB = 0$$

with $B(T) = F - \max\{F - V(T); 0\}$. Alternatively, one can apply the conditions (5.10) to get

$$S_t + rS_vV + 0.5S_{vv}V^2\sigma^2 - rS = 0$$

with $S(T) = \max\{V(T) - F; 0\}$. Obviously, these results yield to formulae for the stock and bond price which were first derived by Merton (1974):

$$\begin{aligned} S(t) &= V(t)\mathcal{N}(d_1) - Fe^{r(T-t)}\mathcal{N}(d_2), \qquad (5.11)\\ B(t) &= Fe^{r(T-t)}\mathcal{N}(d_2) + V(t)\mathcal{N}(-d_1) \end{aligned}$$

with

$$d_1 = \frac{\ln(\frac{V(t)}{F}) + (r + 0.5\sigma^2)(T-t)}{\sigma\sqrt{T-t}}, \qquad d_2 = d_1 - \sigma\sqrt{T-t}.$$

It is worth pointing out that no preferences enter these formulae.

We end with some important remarks: Firm value is only an artificial quantity which is equal to the sum of the firm's liabilities. Besides, the volatility of firm value is not observable. However, given that we can observe the current value of equity $S(t)$ and its volatility σ_S, we obtain two equations for σ and $V(t)$. The first equation is given by (5.11). Combining (5.7) and (5.8), we conclude that the volatility of firm value is related to the volatility of equity by the following equation:

$$S\sigma_S = S_v V\sigma,$$

where $S_v = \mathcal{N}(d_1)$. Hence, we are able to compute σ and $V(t)$. The following table provides numerical examples illustrating our findings.

Table 5.1: Firm value $V(t)$ and volatility σ of the firm. The values are reported for different times to maturity $T-t$, different stock volatilities σ_S, and different face values F as well as $S = 100$ and $r = 0.05$.

$T-t$	1	2	3	4	5	10	
$V(t)$	176.10	172.39	168.84	165.43	162.15	147.53	$F = 80$
σ	0.1704	0.1741	0.1779	0.1819	0.18615	0.2079	$\sigma_S = 0.3$
$V(t)$	176.08	172.08	167.79	163.33	158.82	138.75	$F = 80$
σ	0.2842	0.2938	0.3059	0.3189	0.3318	0.3878	$\sigma_S = 0.5$
$V(t)$	147.56	145.24	143.03	140.92	138.89	129.93	$F = 50$
σ	0.2033	0.2066	0.2098	0.2131	0.2164	0.2331	$\sigma_S = 0.3$
$V(t)$	147.56	145.13	142.58	139.94	137.27	125.26	$F = 50$
σ	0.3389	0.3460	0.3549	0.3647	0.3744	0.4164	$\sigma_S = 0.5$

Finally, we wish to stress that similar results hold for more sophisticated firm value models.

5.2.3 Black-Cox Model

In contrast to Black/Scholes (1973) and Merton (1974), Black/Cox (1976) consider the impact of safety covenants on the value of the firm's securities. This contractual provision gives the bondholders the right to trigger default when the value of the firm touches a lower bound

$$L(t) = k \cdot e^{-\kappa(T_B - t)}$$

with constants $k, \kappa > 0$. Let $\tau := \inf\{t \geq 0 : V(t) = L(t)\}$ be the corresponding stopping time. If default occurs during the lifetime of the bond, the bondholders immediately obtain the ownership of the firm's entire assets. Otherwise, the terminal value of the bond is identical to the value in Merton's model. Hence, the defaultable bond corresponds to a portfolio of barrier derivatives with curved boundary L and maturity equal to the maturity T_B of the bond. More precisely, the corporate bond has the following value at T_B

$$B(T_B, V(T_B)) = F \cdot P_{DO}(T_B) - \max\{F - V(T_B), 0\}_{DO} + H(T_B),$$

where P_{DO} denotes a down-and-out bond with knock-out barrier L and H a cash-at-hit option which will pay the value of the firm if the barrier is touched, i.e. $H(\tau) = k \cdot e^{-\kappa(T_B - \tau)}$ given that $\tau < T_B$. The second term stands for a down-and-out put. If the down-and-out properties are neglected, the first two terms are equal to the final bond value in Merton's model. However, the third term is new. The reader should be aware of the difference between a down-and-in bond and a cash-at-hit option. Whereas the first leads to a constant payment *at maturity* if a barrier is touched during the lifetime of the down-and-in bond, the second *immediately* pays a constant amount which leads to the final value

$$H(T_B) = \begin{cases} k \cdot e^{(r-\kappa)\cdot(T_B - \tau)} & if\ \tau < T_B, \\ 0 & if\ \tau > T_B. \end{cases}$$

As a consequence, a cash-at-hit option only corresponds to a down-and-in bond if $r = \kappa$, i.e. if the growth rate of the boundary equals the compounding rate of the money market account. In this special case, the stochastic payment date of the cash-at-hit option is irrelevant. We call $L(t) = ke^{-r(T_B - t)}$ a *discounted barrier.*

For $k \leq F$ the value of the bond at time $t \in [0, \min\{T_B, \tau\}]$ is given by[4]

$$\begin{aligned} B(t) = {} & Fe^{-r(T_B - t)} \Big[\mathcal{N}(z_1(t)) - y^{2\theta - 2}(t)\mathcal{N}(z_2(t))\Big] \\ & + V(t)\Big[\mathcal{N}(-z_3(t)) + y^{2\theta}(t)\mathcal{N}(z_4(t))\Big], \end{aligned} \tag{5.12}$$

where

$$z_{1/3}(t) = \frac{\ln(\frac{V(t)}{F}) + (r \mp 0.5\sigma^2)(T_B - t)}{\sigma\sqrt{T_B - t}},$$

[4] In Black/Cox (1976) there is a typing error in their formula (8) for the bond value. In their notation, the seventh term of their formula (8) should be $y^{\theta+\eta}\mathcal{N}(z_7)$ instead of $y^{\theta-\eta}\mathcal{N}(z_7)$. Applying Proposition 4.3 it is easy to verify that it has to be $+\eta$. As a consequence, the last four terms in (8) cancel each other out if the dividend yield is equal to zero. Note that we have tried to adopt the notation of Black and Cox, but we have changed the sign of z_3.

$$z_{2/4}(t) = \frac{\ln(\frac{V(t)}{F}) + 2\ln(y(t)) + (r \mp 0.5\sigma^2)(T_B - t)}{\sigma\sqrt{T_B - t}},$$

$y(t) = ke^{-\kappa(T_B - t)}/V(t)$, and $\theta = (r - \kappa + 0.5\sigma^2)/\sigma^2$. As in Section 5.2, the elasticity of the bond plays an important role for the optimal portfolio process. Therefore, we compute the derivative of the bond value with respect to firm value. Rearranging (5.12) leads to a representation which proves useful for these matters:

$$B(t) = \underbrace{Fe^{-r(T_B-t)} - \left[Fe^{-r(T_B-t)}\mathcal{N}(-z_1(t)) - V(t)\mathcal{N}(-z_3(t))\right]}_{\text{Merton's bond price}}$$
$$+y^{2\theta-2}(t)\Big[\underbrace{V(t)y^2(t)\mathcal{N}(z_4(t)) - Fe^{-r(T_B-t)}\mathcal{N}(z_2(t))}_{=C(t)}\Big].$$

The bond price equals the sum of the bond price in Merton's model, which itself is the difference between a put price and the price of a riskless bond, and a correction term which is equal to a number of calls with fictitious underlying $V(t)y^2(t)$. The corresponding call value is denoted by C.

Hence, the derivative of the bond with respect to firm value is given by

$$B_v(t) = \mathcal{N}(-z_3(t)) - y^{2\theta}(t)\left[\tfrac{2\theta-2}{V(t)y^2(t)}C(t) + \mathcal{N}(z_4(t))\right].$$

The first term corresponds to the derivative in Merton's model, the second term is a consequence of the safety covenant. It can be shown that this derivative is positive and smaller than one.

If the lower bound is not touched, the value of equity at the bond's maturity T_B is equal to the value in Merton's model, otherwise the value equals zero. Hence, we get

$$S(t, V(t)) = \max\{V(t) - F, 0\}_{DO},$$

i.e. the stock price equals the price of a down-and-out call. Since $0 \leq B_v \leq 1$, the *accounting equation* $V = S + B$ leads to $0 \leq S_v \leq 1$.

As in Merton's model the bond price corresponds to a portfolio of derivatives. Furthermore, equity is modeled as a single derivative. Hence, an investor will try to track his optimal elasticity. However, if the firm is triggered to default before the investment horizon T, the assets of the firm are handed over to the bondholders. We assume that, once default has occurred, the investor can only put his wealth into the money market account. For instance, this is the case when the investor is focused on a bond investment. Therefore, the only attainable elasticity after default is equal to zero.

Up to default the investor can put funds into securities of the firm. As before, firm value is modeled by

$$dV(t) = V(t)\Big[\alpha dt + \sigma dW(t)\Big], \quad V(0) = v_0.$$

We assume that zero bonds with maturity T_B are issued, where T_B is greater than the investment horizon T. Recall that default of the firm is triggered when a lower bound $L(t) = ke^{-\kappa(T_B - t)}$ is reached. For the sake of simplicity, suppose that $k < v_0$. The wealth equation of the investor reads as follows:

$$dX(t) = X(t)\Big[(r + \lambda\varepsilon(t))dt + \sigma\varepsilon(t)dW(t)\Big],$$

where ε denotes the elasticity of the portfolio with respect to firm value and $\lambda := \alpha - r$. Then the portfolio problem of an investor who maximizes utility from terminal wealth at time T with respect to a power utility function $U(x) = \frac{1}{\gamma}x^{\gamma}$, $\gamma \in (-\infty, 0) \cup (0, 1)$, reads

$$\max_{\varepsilon(\cdot)\in\mathcal{A}'(0,x_0)} \mathrm{E}\Big(\tfrac{1}{\gamma}\Big[X^{\varepsilon}(T)\Big]^{\gamma}\Big) \tag{5.13}$$

with

$$dX^{\varepsilon}(t) = X^{\varepsilon}(t)\Big[(r + \varepsilon(t)\lambda)dt + \varepsilon(t)\sigma dW(t)\Big],$$
$$X^{\varepsilon}(0) = x_0,$$

and

$$\mathcal{A}'(0, x_0) := \Big\{\varepsilon(\cdot) \in \mathcal{A}^*(0, x_0) \,:\, \varepsilon(t) = 0 \,\forall\, t \in [\tau, T]\Big\},$$

where $\mathcal{A}^*(0, x_0) := \Big\{\varepsilon(\cdot) \in \mathcal{A}(0, x_0) \,:\, X^{\varepsilon}(t) \geq 0 \text{ for } t \in [0, T]\Big\}$. However, Corollary 1.1 and the specific form of every coefficient in the wealth equation indeed guarantees the positivity of $X^{\varepsilon}(t)$. Therefore, we obtain $\mathcal{A}^*(0, x_0) = \mathcal{A}(0, x_0)$.

We face a two-dimensional control problem with state process (X, V) and

$$\Lambda(t, x, v, \varepsilon) = (x(r + \lambda\varepsilon), v\alpha)',$$
$$\Sigma(t, x, v, \varepsilon) = (x\varepsilon\sigma, v\sigma)',$$
$$\Sigma^*(t, x, v, \varepsilon) = \begin{pmatrix} x^2\varepsilon^2\sigma^2 & vx\sigma^2\varepsilon \\ vx\sigma^2\varepsilon & v^2\sigma^2 \end{pmatrix},$$
$$A^{\varepsilon}G(t, x, v) = G_t + 0.5(x^2\varepsilon^2\sigma^2 G_{xx} + 2vx\varepsilon\sigma^2 G_{xv} + \sigma^2 v^2 G_{vv}) + x(r + \lambda\varepsilon)G_x + v\alpha G_v.$$

Hence, the following Hamilton-Jacobi-Bellman equation (HJB) needs to be solved:

$$\sup_{\varepsilon} A^{\varepsilon} G(t,x,v) = 0 \tag{5.14}$$

with terminal condition $G(T,x,v) = \frac{1}{\gamma}x^{\gamma}$. Since we assume that once default is triggered the investor wishes to invest his funds in the money market account (or is forced to do that), after default the value function reads

$$G(t,x,v) = \mathrm{E}^{t,x,v}\Big(\tfrac{1}{\gamma}X(T)^{\gamma}\Big) = \tfrac{1}{\gamma}x^{\gamma}\exp\Big(\gamma(T-t)r\Big), \quad t \geq \tau,$$

which leads to an additional boundary condition $G(t,x,L(t)) = \frac{1}{\gamma}x^{\gamma}\exp(\gamma(T-t)r)$. Pointwise maximization over ε in (5.14) leads to the candidate for the optimal elasticity

$$\varepsilon^* = -\frac{\lambda}{\sigma^2}\frac{G_x}{xG_{xx}} - \frac{v}{x}\frac{G_{xv}}{G_{xx}},$$

where $G_{xx} < 0$ is assumed to be satisfied. Plugging ε^* into the HJB leads to the PDE

$$\begin{aligned} 0 = {} & G_t G_{xx} + xrG_xG_{xx} - 0.5\tfrac{\lambda^2}{\sigma^2}G_x^2 + v\Big(\alpha G_v G_{xx} - \lambda G_{xv}G_x\Big) \\ & +0.5v^2\sigma^2\Big(G_{vv}G_{xx} - G_{xv}^2\Big) \end{aligned}$$

for the value function. Applying the separation ansatz

$$G(t,x,v) = \tfrac{1}{\gamma}x^{\gamma}(f(t,v))^{1-\gamma}$$

with $f(T,v) = 1$ and $f(t,L(t)) = \exp(\frac{\gamma}{1-\gamma}(T-t)r)$ yields

$$0 = f_t + \underbrace{\Big(\tfrac{\gamma}{1-\gamma}r + 0.5\tfrac{\gamma}{(1-\gamma)^2}\tfrac{\lambda^2}{\sigma^2}\Big)}_{=:-\tilde{r}}f + \underbrace{\Big(\tfrac{\gamma}{1-\gamma}\lambda + \alpha\Big)}_{=:\tilde{\mu}}vf_v + 0.5v^2\sigma^2 f_{vv}.$$

To solve this PDE for f, we apply a Feynman-Kac representation theorem for boundary value problems. The proof is presented at the end of the subsection.

Proposition 5.3 (Feynman-Kac for a Boundary Value Problem) *Consider the boundary value problem*

$$0 = f_t - \tilde{r}f + \tilde{\mu}vf_v + 0.5v^2\sigma^2 f_{vv} \tag{5.15}$$

with $f(t,L(t)) = \psi_1(t)$ and $f(T,v) = \psi_2(v)$, where ψ_1, ψ_2, and L are non-negative, continuous, and deterministic functions and $\tilde{E}^{t,v}(\psi_2(V(T)) < +\infty$. Define the stopping time $\tau := \inf\{s \geq 0 : V(s) = L(s)\}$ and suppose that

$$\tilde{f}(t,v) := \tilde{\mathrm{E}}^{t,v}\left(1_{(\tau<T)}e^{-\tilde{r}(\tau-t)}\psi_1(\tau) + 1_{(\tau\geq T)}e^{-\tilde{r}(T-t)}\psi_2(V(T))\right),$$

$0 \leq t < \tau \wedge T$, $v > L(t)$, with

$$dV(t) = V(t)\left[\tilde{\mu}dt + \sigma d\tilde{W}(t)\right]$$

is continuous and an element of $C^{1,2}([0,T) \times I\!R^+)$. Then $\tilde{f}$ meets the PDE (5.15).

Due to this proposition we first compute the following expectation:

$$\begin{aligned} f(t,v) &= \tilde{\mathrm{E}}^{t,v}\left(1_{(\tau<T)} \cdot e^{\frac{\gamma}{1-\gamma}r(T-\tau)}e^{-\tilde{r}(\tau-t)} + 1_{(\tau\geq T)} \cdot e^{-\tilde{r}(T-t)}\right) \\ &= \tilde{\mathrm{E}}^{t,v}\left(1_{(\tau<T)} \cdot e^{c\tau}\right)e^{\frac{\gamma}{1-\gamma}rT+\tilde{r}t} + \tilde{P}^{t,v}(\tau \geq T)e^{-\tilde{r}(T-t)}, \end{aligned}$$

$c := 0.5\frac{\gamma}{(1-\gamma)^2}\frac{\lambda^2}{\sigma^2}$, where under $\tilde{P}$ the value of the firm V follows the dynamics

$$dV = V\left[\tilde{\mu}dt + \sigma d\tilde{W}\right]$$

and $\tilde{W}$ is a Brownian motion under $\tilde{P}$. Note that $f > 0$ and thus $G_{xx} < 0$ is met. Consider

$$\begin{aligned} \{V(s) = L(s)\} &= \left\{v_0 \exp\left((\tilde{\mu} - 0.5\sigma^2)s + \sigma\tilde{W}(s)\right) = k\exp(-\kappa(T_B - s))\right\} \\ &= \left\{\tfrac{\tilde{\mu}-0.5\sigma^2-\kappa}{\sigma}s + \tilde{W}(s) = \tfrac{1}{\sigma}\left(\ln(\tfrac{k}{v_0}) - \kappa T_B\right)\right\} \\ &= \Big\{\underbrace{-\tfrac{\tilde{\mu}-0.5\sigma^2-\kappa}{\sigma}}_{=:\theta}s - \tilde{W}(s) = \underbrace{\tfrac{1}{\sigma}\left(\ln(\tfrac{v_0}{k}) + \kappa T_B\right)}_{=:\tilde{H}>0}\Big\}. \end{aligned}$$

Therefore, τ is the first hitting time of a Brownian motion with drift and hit level $\tilde{H}$. Its distribution function reads[5]

$$F_\tau(u) = \mathcal{N}\left(\tfrac{\theta u-\tilde{H}}{\sqrt{u}}\right) + e^{2\theta\tilde{H}} \cdot \mathcal{N}\left(\tfrac{-\theta u-\tilde{H}}{\sqrt{u}}\right).$$

Hence, τ has the following density:

$$f_\tau(u) = \frac{\tilde{H}}{u\sqrt{2\pi u}} \cdot \exp\left(-\frac{(\tilde{H}-\theta u)^2}{2u}\right).$$

Note that given some initial value (t, v_t) with $v_t > L(t)$ the above event can be rewritten as

[5] See e.g. Elliott/Kopp (1999), p. 176.

$$
\begin{aligned}
&\{V(s) = L(s)\} \\
&= \Big\{v_t \exp\Big((\tilde{\mu} - 0.5\sigma^2)(s-t) + \sigma\tilde{W}(s-t)\Big) = k\exp(-\kappa(T_B - s))\Big\} \\
&= \Big\{\tfrac{\tilde{\mu}-0.5\sigma^2-\kappa}{\sigma}(s-t) + \tilde{W}(s-t) = \tfrac{1}{\sigma}\Big(\ln(\tfrac{k}{v_t}) - \kappa(T_B - t)\Big)\Big\} \\
&= \Big\{\underbrace{-\tfrac{\tilde{\mu}-0.5\sigma^2-\kappa}{\sigma}}_{=\theta}(s-t) - \tilde{W}(s-t) = \underbrace{\tfrac{1}{\sigma}\Big(\ln(\tfrac{v_t}{k}) + \kappa(T_B - t)\Big)}_{=:\tilde{H}^{t,v_t}>0}\Big\},
\end{aligned}
$$

i.e. $\tilde{H}^{t,v_t}$ is a function of t and v_t. Fix $v > L(t)$. If $\theta^2 \geq 2c$, we obtain

$$
\begin{aligned}
&\tilde{\mathrm{E}}^{t,v}\Big(1_{(\tau<T)} \cdot e^{c\tau}\Big) = \int_0^{T-t} f_\tau(u) \cdot e^{c(u+t)}\,du \\
&= e^{ct}\int_0^{T-t} \tfrac{\tilde{H}^{t,v}}{u\sqrt{2\pi u}} \cdot \exp\Big(-\tfrac{(\tilde{H}^{t,v}-\theta u)^2}{2u}\Big) \cdot e^{cu}\,du \\
&= e^{ct+\tilde{H}^{t,v}(\theta-\sqrt{\theta^2-2c})}\int_0^{T-t} \tfrac{\tilde{H}^{t,v}}{u\sqrt{2\pi u}} \cdot \exp\Big(-\tfrac{(\tilde{H}^{t,v}-\sqrt{\theta^2-2c}\,u)^2}{2u}\Big)\,du \\
&= e^{ct+\tilde{H}^{t,v}(\theta-\sqrt{\theta^2-2c})} \\
&\quad\cdot\Big(\mathcal{N}\Big(\tfrac{\sqrt{\theta^2-2c}\cdot(T-t)-\tilde{H}^{t,v}}{\sqrt{T-t}}\Big) + e^{2\sqrt{\theta^2-2c}\cdot\tilde{H}^{t,v}} \cdot \mathcal{N}\Big(\tfrac{-\sqrt{\theta^2-2c}\cdot(T-t)-\tilde{H}^{t,v}}{\sqrt{T-t}}\Big)\Big).
\end{aligned}
$$

On the other hand, we get

$$
\tilde{P}^{t,v}(\tau \geq T) = 1 - F_\tau(T-t) = \mathcal{N}\Big(\tfrac{-\theta(T-t)+\tilde{H}^{t,v}}{\sqrt{T-t}}\Big) - e^{2\theta\tilde{H}^{t,v}} \cdot \mathcal{N}\Big(\tfrac{-\theta(T-t)-\tilde{H}^{t,v}}{\sqrt{T-t}}\Big).
$$

Combining both results, we arrive at

$$
\begin{aligned}
&f(t,v) \\
&= e^{\frac{\gamma}{1-\gamma}rT+\tilde{r}t} \cdot \tilde{\mathrm{E}}^{t,v}\Big(1_{(\tau<T)} \cdot e^{c\tau}\Big) + e^{-\tilde{r}(T-t)} \cdot \tilde{P}^{t,v}(\tau \geq T) \qquad (5.16) \\
&= e^{\frac{\gamma}{1-\gamma}r(T-t)+\tilde{H}^{t,v}(\theta-\sqrt{\theta^2-2c})} \\
&\quad\cdot\Big(\mathcal{N}\Big(\tfrac{\sqrt{\theta^2-2c}\cdot(T-t)-\tilde{H}^{t,v}}{\sqrt{T-t}}\Big) + e^{2\sqrt{\theta^2-2c}\cdot\tilde{H}^{t,v}} \cdot \mathcal{N}\Big(\tfrac{-\sqrt{\theta^2-2c}\cdot(T-t)-\tilde{H}^{t,v}}{\sqrt{T-t}}\Big)\Big) \\
&\quad+ e^{-\tilde{r}(T-t)} \cdot \Big(\mathcal{N}\Big(\tfrac{-\theta(T-t)+\tilde{H}^{t,v}}{\sqrt{T-t}}\Big) - e^{2\theta\tilde{H}^{t,v}} \cdot \mathcal{N}\Big(\tfrac{-\theta(T-t)-\tilde{H}^{t,v}}{\sqrt{T-t}}\Big)\Big).
\end{aligned}
$$

Defining the first hitting time $\tilde{\tau} := \inf\{s \geq 0 : \sqrt{\theta^2 - 2c}\cdot s + \tilde{W}(s) = \tilde{H}\}$, this result can be rewritten as

$$
\begin{aligned}
f(t,v) = e^{\frac{\gamma}{1-\gamma}r(T-t)}\Big(&e^{\tilde{H}^{t,v}(\theta-\sqrt{\theta^2-2c})} \cdot \tilde{P}^{t,v}(\tilde{\tau} \leq T) \\
&+ e^{0.5\frac{\gamma}{(1-\gamma)^2}\frac{\lambda^2}{\sigma^2}(T-t)} \cdot \tilde{P}^{t,v}(\tau \geq T)\Big).
\end{aligned}
$$

Note that in the special case $\kappa = 0$ we get

$$\begin{aligned}
&f(t,v)\\
&= e^{\frac{\gamma}{1-\gamma}r(T-t)}\left(\tfrac{v}{k}\right)^{\sigma^{-1}(\theta-\sqrt{\theta^2-2c})}\\
&\quad\cdot\left(\mathcal{N}\left(\tfrac{\sqrt{\theta^2-2c}\cdot(T-t)-\tilde{H}^{t,v}}{\sqrt{T-t}}\right)+\left(\tfrac{v}{k}\right)^{2\sigma^{-1}\sqrt{\theta^2-2c}}\cdot\mathcal{N}\left(\tfrac{-\sqrt{\theta^2-2c}\cdot(T-t)-\tilde{H}^{t,v}}{\sqrt{T-t}}\right)\right)\\
&\quad+e^{-\tilde{r}(T-t)}\cdot\left(\mathcal{N}\left(\tfrac{-\theta(T-t)+\tilde{H}^{t,v}}{\sqrt{T-t}}\right)-\left(\tfrac{v}{k}\right)^{2\theta\sigma^{-1}}\cdot\mathcal{N}\left(\tfrac{-\theta(T-t)-\tilde{H}^{t,v}}{\sqrt{T-t}}\right)\right).
\end{aligned}$$

For the validity of these results it is crucial that $\theta^2 - 2c \geq 0$. It is straightforward to verify that

$$\theta^2 - 2c = \tfrac{\gamma}{1-\gamma}\tfrac{\alpha-r}{\sigma^2}\cdot\left(2(\alpha-0.5\sigma^2-\kappa)-\alpha+r\right)+\tfrac{(\alpha-0.5\sigma^2-\kappa)^2}{\sigma^2}.$$

To this end, for $0 < \gamma < 1$ we make the following assumptions:

$$\alpha \geq r, \tag{5.17}$$

$$\alpha + r - \sigma^2 - 2\kappa \geq 0, \tag{5.18}$$

$$\alpha - 0.5\sigma^2 - \kappa \geq 0. \tag{5.19}$$

Assumptions (5.17) and (5.18) are sufficient to guarantee that $\theta^2 - 2c \geq 0$. Further, assumptions (5.17) and (5.19) imply $\theta \leq 0$, which will be needed later on. The candidate for the optimal elasticity reads

$$\varepsilon^*(t) = \frac{1}{1-\gamma}\frac{\lambda}{\sigma^2} + \frac{V(t)\cdot f_v(t,V(t))}{f(t,V(t))},$$

where with $w := \sqrt{\theta^2 - 2c}$ we get

$$\begin{aligned}
&v\cdot f_v(t,v)\\
&= e^{\frac{\gamma}{1-\gamma}r(T-t)+\tilde{H}^{t,v}(\theta-w)}\cdot\tfrac{\theta-w}{\sigma}\cdot\left\{\mathcal{N}\left(\tfrac{w\cdot(T-t)-\tilde{H}^{t,v}}{\sqrt{T-t}}\right)\right.\\
&\qquad\qquad\left.+e^{2w\tilde{H}^{t,v}}\mathcal{N}\left(\tfrac{-w(T-t)-\tilde{H}^{t,v}}{\sqrt{T-t}}\right)\right\}\\
&\quad+e^{\frac{\gamma}{1-\gamma}r(T-t)+\tilde{H}^{t,v}(\theta-w)}\cdot\left[n\left(\tfrac{w(T-t)-\tilde{H}^{t,v}}{\sqrt{T-t}}\right)\cdot\tfrac{-1}{\sigma\sqrt{T-t}}\right.\\
&\qquad\qquad+2\tfrac{w}{\sigma}e^{2w\tilde{H}^{t,v}}\cdot\mathcal{N}\left(\tfrac{-w(T-t)-\tilde{H}^{t,v}}{\sqrt{T-t}}\right)\\
&\qquad\qquad\left.+e^{2w\tilde{H}^{t,v}}\cdot n\left(\tfrac{-w(T-t)-\tilde{H}^{t,v}}{\sqrt{T-t}}\right)\cdot\tfrac{-1}{\sigma\sqrt{T-t}}\right]\\
&\quad+e^{-\tilde{r}(T-t)}\cdot\left[n\left(\tfrac{-\theta(T-t)+\tilde{H}^{t,v}}{\sqrt{T-t}}\right)\cdot\tfrac{1}{\sigma\sqrt{T-t}}-e^{2\theta\tilde{H}^{t,v}}\cdot 2\tfrac{\theta}{\sigma}\cdot\mathcal{N}\left(\tfrac{-\theta(T-t)-\tilde{H}^{t,v}}{\sqrt{T-t}}\right)\right.
\end{aligned}$$

$$
\begin{aligned}
&\qquad -e^{2\theta\tilde{H}^{t,v}} \cdot n\left(\frac{-\theta(T-t)-\tilde{H}^{t,v}}{\sqrt{T-t}}\right) \cdot \frac{-1}{\sigma\sqrt{T-t}}\Bigg] \\
&= e^{\frac{\gamma}{1-\gamma}r(T-t)+\tilde{H}^{t,v}(\theta-w)} \cdot \left\{\frac{\theta-w}{\sigma}\mathcal{N}\left(\frac{w\cdot(T-t)-\tilde{H}^{t,v}}{\sqrt{T-t}}\right)\right. \\
&\qquad\qquad \left. +\frac{\theta+w}{\sigma}e^{2w\tilde{H}^{t,v}}\mathcal{N}\left(\frac{-w(T-t)-\tilde{H}^{t,v}}{\sqrt{T-t}}\right)\right\} \\
&\quad -2\frac{\theta}{\sigma}\cdot e^{-\tilde{r}(T-t)+2\theta\tilde{H}^{t,v}} \cdot \mathcal{N}\left(\frac{-\theta(T-t)-\tilde{H}^{t,v}}{\sqrt{T-t}}\right).
\end{aligned}
\tag{5.20}
$$

To verify that G is the value function of the problem and that ε^* is the optimal elasticity, we now show that ε^* is bounded. First note that due to (5.16) the function f is positive and bounded away from zero because

$$
\begin{aligned}
f(t,v) &\geq \min\left\{e^{\frac{\gamma}{1-\gamma}rT+\tilde{r}t}; e^{-\tilde{r}(T-t)}\right\} \cdot \mathrm{E}\Big(1_{(\tau<T)}\min\{e^{cT},1\}\Big) \\
&\quad + \min\left\{e^{\frac{\gamma}{1-\gamma}rT+\tilde{r}t}; e^{-\tilde{r}(T-t)}\right\} \cdot \mathrm{E}\Big(1_{(\tau\geq T)}\min\{e^{cT},1\}\Big) \\
&\geq \min\left\{\min_{0\leq t\leq T} e^{\frac{\gamma}{1-\gamma}rT+\tilde{r}t}; \min_{0\leq t\leq T} e^{-\tilde{r}(T-t)}\right\} \cdot \min\{e^{cT},1\} = \mathit{const.} > 0.
\end{aligned}
$$

Consequently, ε^* is bounded if vf_v is bounded. Applying the triangle inequality and adding a suitable value of the cumulative normal distribution function, we get the following estimate

$$
\begin{aligned}
&|v\cdot f_v(t,v)| \\
&\quad \leq e^{\frac{\gamma}{1-\gamma}r(T-t)+\tilde{H}^{t,v}(\theta-w)} \cdot \frac{|\theta|+w}{\sigma} \cdot \left\{\mathcal{N}\left(\frac{w\cdot(T-t)-\tilde{H}^{t,v}}{\sqrt{T-t}}\right)\right. \\
&\qquad\qquad \left. +e^{2w\tilde{H}^{t,v}}\mathcal{N}\left(\frac{-w(T-t)-\tilde{H}^{t,v}}{\sqrt{T-t}}\right)\right\} \\
&\qquad +2\frac{|\theta|}{\sigma}\cdot e^{-\tilde{r}(T-t)}\left\{\mathcal{N}\left(\frac{\theta(T-t)-\tilde{H}^{t,v}}{\sqrt{T-t}}\right) + e^{2\theta\tilde{H}^{t,v}} \cdot \mathcal{N}\left(\frac{-\theta(T-t)-\tilde{H}^{t,v}}{\sqrt{T-t}}\right)\right\}. \\
&\quad \leq e^{\frac{\gamma}{1-\gamma}r(T-t)+\tilde{H}^{t,v}(\theta-w)} \cdot \frac{|\theta|+w}{\sigma} + 2\frac{|\theta|}{\sigma}\cdot e^{-\tilde{r}(T-t)}
\end{aligned}
\tag{5.21}
$$

First notice that $\tilde{H}^{t,v} > 0$. Hence, for $0 < \gamma < 1$ the term $e^{\tilde{H}^{t,v}(\theta-w)}$ is bounded because $|\theta| \geq w$ and, by assumption (5.19), $\theta \leq 0$. On the other hand, for $\gamma < 0$ it is bounded because $w \geq \theta$. Consequently, (5.21) and therefore ε^* are bounded. We wish to stress that our state process (X,V) meets the Lipschitz and growth conditions given that we first limit ourselves to bounded admissible controls. Further, if $\gamma > 0$, the function G fulfils the polynomial growth condition on the value function because f is bounded. Hence, Theorem 1.3 can be applied. Therefore, G is the value function of the problem and ε^* is the optimal control even if unbounded admissible controls are considered.

In the case $\gamma < 0$ the polynomial growth conditions (1.9) and (1.11) are not satisfied. Nevertheless, as in Corollary 1.2 it is sufficient to prove that for bounded admissible controls the condition (1.25) is met. Then our candidate ε^* is optimal among all admissible controls. Given $\alpha > 1$, $\gamma < 0$, and a bounded admissible control ε we obtain

$$\mathrm{E}\Big(|G(t, X^\varepsilon(t))|^\alpha\Big) = \mathrm{E}\Big(\tfrac{1}{\gamma}\Big[X^\varepsilon(t)\Big]^{\gamma\alpha}\Big[f(t, V(t))\Big]^{\alpha(1-\gamma)}\Big) \leq K_1 \,\mathrm{E}\Big(\Big[X^\varepsilon(t)\Big]^{\gamma\alpha}\Big),$$

where $K_1 > 0$ is a suitable constant. With $\beta := \alpha\gamma$ and a suitable constant $K_2 > 0$ it follows that

$$\begin{aligned}
&\Big[X^\varepsilon(t)\Big]^\beta \\
&= x_0^\beta \exp\left(\beta r t + \beta \int_0^t \lambda\varepsilon(s) - 0.5\sigma^2\varepsilon^2(s)\, ds + \beta\sigma \int_0^t \varepsilon(s)\, dW(s)\right) \\
&\leq K_2 \exp\left(\beta\sigma \int_0^t \varepsilon(s)\, dW(s)\right) \\
&= K_2 \exp\left(0.5\beta^2\sigma^2 \int_0^t \varepsilon^2(s)\, ds - 0.5\beta^2\sigma^2 \int_0^t \varepsilon^2(s)\, ds + \beta\sigma \int_0^t \varepsilon(s)\, dW(s)\right).
\end{aligned}$$

Hence, we arrive at

$$\mathrm{E}\Big(\Big[X^\varepsilon(t)\Big]^\beta\Big) \leq K_2 \exp\left(0.5\beta^2\sigma^2 \int_0^t \varepsilon^2(s)\, ds.\right),$$

which leads to

$$\mathrm{E}\Big(\sup_{t\in[0,T]} |G(t, X^\varepsilon(t))|^\alpha\Big) < \infty$$

and that proves (1.25). The following proposition summarizes our results.

Proposition 5.4 (Optimal Elasticity in the Black-Cox Model)
Consider the portfolio problem (5.13) and the Black-Cox model. Suppose that for $\gamma \in (0,1)$ the assumptions (5.17) - (5.19) are satisfied. Then the optimal elasticity up to default is given by

$$\varepsilon^*(t) = \frac{1}{1-\gamma}\frac{\lambda}{\sigma^2} + \frac{V(t) f_v(t, V(t))}{f(t, V(t))},$$

$0 \leq t < \tau$, *where f is given by (5.16).*

Applying this proposition, we can compute the optimal portfolio processes.

Corollary 5.1 (Optimal Portfolios in the Black-Cox Model)
(i) If the investor whishes to put his wealth into the riskless money market account and into stocks, then the optimal fraction invested in stocks is uniquely determined and equals

$$\pi_S^*(t) = \frac{\varepsilon^*(t)}{\varepsilon_S(t)} = \begin{cases} \frac{1}{1-\gamma}\frac{\lambda}{\sigma^2}\frac{S(t)}{V(t)S_v(t)} + \frac{f_v(t,V(t))\,S(t)}{f(t,V(t))\,S_v(t)}, & \text{if} \quad \tau > t, \\ 0, & \text{if} \quad \tau \leq t, \end{cases}$$

where $S_v(t) = \mathcal{N}(z_3(t)) + y^{2\theta}(t)\left[\frac{2\theta-2}{V(t)y^2(t)}C(t) + \mathcal{N}(z_4(t))\right]$ *and* f *is given by (5.16).*

(ii) If the investor wishes to put his wealth into the riskless money market account and into defaultable bonds, then the optimal fraction invested in defaultable bonds is uniquely determined and equals

$$\pi_B^*(t) = \frac{\varepsilon^*(t)}{\varepsilon_B(t)} = \begin{cases} \frac{1}{1-\gamma}\frac{\lambda}{\sigma^2}\frac{B(t)}{V(t)B_v(t)} + \frac{f_v(t,V(t))\,B(t)}{f(t,V(t))\,B_v(t)}, & \text{if} \quad \tau > t, \\ 0, & \text{if} \quad \tau \leq t, \end{cases}$$

where $B_v(t) = \mathcal{N}(-z_3(t)) - y^{2\theta}(t)\left[\frac{2\theta-2}{V(t)y^2(t)}C(t) + \mathcal{N}(z_4(t))\right]$ *and* f *is given by (5.16).*

(iii) If the investor wishes to allocate his wealth to the riskless money market account, to defaultable bonds, and to stocks, then the optimal portfolio process is not uniquely determined. Every combination of stocks and bonds is optimal if it leads to the optimal elasticity.

We end this section with the proof of Proposition 5.3 and a remark on one of its assumptions.

Proof of Proposition 5.3: We extend a proof of Shreve (1997, p. 215), who deals with up-and-out calls without rebates. For simplicity of notation, we omit all tildes in this proof. Further, let $a \wedge b := \min\{a, b\}$, a, $b \in I\!R$. We begin by showing the following property:

$$\left\{e^{-r(t\wedge\tau)} f(t\wedge\tau, V(t\wedge\tau))\right\}_{t\in[0,T]} \quad \text{is a martingale.} \tag{5.22}$$

Fix $\omega \in \Omega$ and $t \in [0, T]$. First, we consider the case $\tau(\omega) \leq t$. Then, we get

$$\mathrm{E}\Big(1_{(\tau<T)}e^{-r\tau}\psi_1(\tau) + 1_{(\tau\geq T)}e^{-rT}\psi_2(V(T))\Big|\mathcal{F}_t\Big)(\omega) = e^{-r\tau(\omega)}\psi_1(\tau(\omega)).$$

Besides,

$$\begin{aligned}
& f(t \wedge \tau(\omega), V(t \wedge \tau(\omega), \omega)) \\
& \quad = f(\tau(\omega), L(\tau(\omega))) \\
& \quad = \mathrm{E}^{\tau(\omega), L(\tau(\omega))}\left(1_{(\tau<T)} e^{-r(\tau-t)} \psi_1(\tau) + 1_{(\tau \geq T)} e^{-r(T-t)} \psi_2(V(T))\right) \\
& \quad = \psi_1(\tau(\omega)).
\end{aligned}$$

Therefore, we conclude

$$\begin{aligned}
\mathrm{E}\Big(1_{(\tau<T)} e^{-r\tau} \psi_1(\tau) + 1_{(\tau \geq T)} e^{-rT} \psi_2(V(T)) \Big| \mathcal{F}_t\Big)(\omega) \\
= e^{-r(t \wedge \tau(\omega))} f(t \wedge \tau(\omega), V(t \wedge \tau(\omega), \omega)).
\end{aligned}$$

In the case $\tau(\omega) > t$ we obtain, due to the Markov property of V,

$$\begin{aligned}
& \mathrm{E}\Big(1_{(\tau<T)} e^{-r\tau} \psi_1(\tau) + 1_{(\tau \geq T)} e^{-rT} \psi_2(V(T)) \Big| \mathcal{F}_t\Big)(\omega) \\
& \quad = \mathrm{E}^{t, V(t,\omega)}\Big(1_{(\tau<T)} e^{-r\tau} \psi_1(\tau) + 1_{(\tau \geq T)} e^{-rT} \psi_2(V(T)) \Big| \mathcal{F}_t\Big) e^{rt} e^{-rt} \\
& \quad = f(t, V(t,\omega)) e^{-rt} \\
& \quad = f(t \wedge \tau(\omega), V(t \wedge \tau(\omega), \omega)) e^{-r(t \wedge \tau(\omega))}.
\end{aligned}$$

Hence, in both cases we arrive at

$$e^{-r(t \wedge \tau)} f(t \wedge \tau, V(t \wedge \tau)) = \mathrm{E}\Big(1_{(\tau<T)} e^{-r\tau} \psi_1(\tau) + 1_{(\tau \geq T)} e^{-rT} \psi_2(V(T)) \Big| \mathcal{F}_t\Big).$$

Since the argument of the conditional expectation on the right-hand side is integrable, the conditional expectation is a martingale and the property (5.22) is proved.

Applying Ito's formula, we obtain

$$\begin{aligned}
d\Big(e^{-rt} f(t, V(t)\Big) &= e^{-rt} df(t, V(t)) + f(t, V(t)) de^{-rt} \\
&= e^{-rt}\Big[f_t dt + f_v dV + 0.5 f_{vv} d < V >\Big] - f r e^{-rt} dt \\
&= e^{-rt}\Big\{f_t dt + f_v V[\mu dt + \sigma dW] + 0.5 f_{vv} \sigma^2 V^2 dt - r f dt\Big\} \\
&= e^{-rt}\Big\{\Big[f_t + \mu V f_v + 0.5 \sigma^2 V^2 f_{vv} - r f\Big] dt + \sigma V f_v dW\Big\},
\end{aligned}$$

where after the second "=" we have omitted the arguments of f, its derivatives, and the stochastic processes for notational convenience. Integrating from 0 to $t \wedge \tau$ leads to

$$
\begin{aligned}
&e^{-r(t\wedge\tau)} f(t\wedge\tau, V(t\wedge\tau)) \\
&\quad = f(0, V(0)) \\
&\quad\quad + \int_0^{t\wedge\tau} e^{-ru} \Big[f_t(u, V(u)) + \mu V(u) f_v(u, V(u)) + 0.5\sigma^2 V^2(u) f_{vv}(u, V(u)) \\
&\quad\quad\quad - r f(u, V(u)) \Big] \, du \\
&\quad\quad + \int_0^{t\wedge\tau} e^{-ru} \sigma V(u) f_v(u, V(u)) \, dW(u)
\end{aligned}
$$

The stopped Ito integral is a martingale. Since $\left\{e^{-r(t\wedge\tau)} f(t\wedge\tau, V(t\wedge\tau))\right\}_{t\in[0,T]}$ is also a martingale, the Lebesgue integral has to be a martingale. According to Ito's martingale representation theorem[6] and the uniqueness of the representation of an Ito process, this leads to

$$
f_t(u, V(u)) + \mu V(u) f_v(u, V(u)) + 0.5\sigma^2 V^2(u) f_{vv}(u, V(u)) - r f(u, V(u)) = 0
$$

for all $0 \leq u \leq t \wedge \tau$. Hence, f meets the PDE (5.15). □

Remark.

It is possible to get rid of the assumption that $f \in C^{1,2}([0,T) \times I\!R^+)$. The idea is to show that under certain conditions the boundary value problem has a unique solution that belongs to $C^{1,2}([0,T) \times I\!R^+)$ by transforming the Black-Scholes PDE into the heat equation. In a second step proof has to be provided that this solution equals f. Nevertheless, for our application we need to calculate the expectation in any case. Therefore, we can directly verify that the solution is indeed a function of $C^{1,2}([0,T) \times I\!R^+)$.

Another way would be to prove that the expectation itself is a function of $C^{1,2}([0,T) \times I\!R^+)$. This proof heavily relies on the joint distribution of V and τ and one would exploit the fact that a joint density exists. Define

$$
\begin{aligned}
Y(s) &:= (\underbrace{\tilde{\mu} - 0.5\sigma^2 - \kappa}_{=:\theta_2})(s-t) + \sigma \tilde{W}(s-t), \\
\tilde{H}_2^{t,v} &:= \ln(k/v) - \kappa(T_B - t), \\
m_Y(T) &:= \min_{s\in[t,T]} Y(s), \\
g^{t,v}(y,m) &:= \tfrac{-2(2m-y)}{\sigma^3 (T-t)^{1.5}} \cdot e^{2\theta_2 m \sigma^{-2}} \cdot n\Big(\tfrac{-y+2m+\theta_2(T-t)}{\sigma\sqrt{T-t}}\Big),
\end{aligned}
$$

[6] See e.g. Korn/Korn(2001), p. 71.

where $n(\cdot)$ denotes the density of a standard normal distribution and $g^{t,v}$ the joint distribution of $Y(T)$ and $m_Y(T)$ given the initial value (t,v).[7] With $h^{t,v}(y) := v \cdot e^y$ we consider

$$
\begin{aligned}
&f(t,v)\\
&= \tilde{\mathrm{E}}^{t,v}\left(1_{(\tau<T)}e^{-\tilde{r}(\tau-t)}\psi_1(\tau) + 1_{(\tau\geq T)}e^{-\tilde{r}(T-t)}\psi_2(V(T))\right)\\
&= \tilde{\mathrm{E}}^{t,v}\left(1_{(\tau<T)}e^{-\tilde{r}(\tau-t)}\psi_1(\tau)\right) + \tilde{\mathrm{E}}^{t,v}\left(1_{(\tau\geq T)}e^{-\tilde{r}(T-t)}\psi_2(V(T))\right)\\
&= \int_0^{T-t} f_\tau(u)e^{-ru}\psi_1(u+t)\,du + e^{-\tilde{r}(T-t)}\tilde{\mathrm{E}}^{t,v}\left(1_{(m_Y(T)>\tilde{H}_2^{t,v})}\psi_2(V(T))\right)\\
&= \int_0^{T-t} f_\tau(u)e^{-ru}\psi_1(u+t)\,du\\
&\quad + e^{-\tilde{r}(T-t)}\int_{\tilde{H}_2^{t,v}}^{0}\int_{\tilde{H}_2^{t,v}}^{+\infty}\psi_2(h^{t,v}(y))g^{t,v}(y,m)\,dy\,dm\\
&= \int_0^{T-t}\frac{\tilde{H}_2^{t,v}}{u\sqrt{2\pi u}}\cdot\exp\left(-\frac{(\tilde{H}_2^{t,v}-\theta u)^2}{2u}\right)e^{-ru}\psi_1(u+t)\,du \quad + e^{-\tilde{r}(T-t)}\\
&\quad\cdot\int_{\tilde{H}_2^{t,v}}^{0}\int_{\tilde{H}_2^{t,v}}^{+\infty}\psi_2(h^{t,v}(y))\frac{-2(2m-y)}{\sigma^3(T-t)^{1.5}}\cdot e^{2\theta_2 m\sigma^{-2}}\cdot n\left(\frac{-y+2m+\theta_2(T-t)}{\sigma\sqrt{T-t}}\right)dy\,dm
\end{aligned}
$$

To verify that f belongs to $C^{1,2}([0,T)\times I\!R^+)$ one needs to show that the last two integrals belong to $C^{1,2}([0,T)\times I\!R^+)$. Clearly, one has to impose additional requirements on ψ_1 and ψ_2. For options which are not path-dependent this is done in Nielsen (1999, pp. 228ff and pp. 370ff).[8] Note that the payoffs of these options do not involve a stopping time τ, i.e. we need only consider the second integral that becomes a single integral. Nevertheless, even then the proof is rather involved.

5.3 From the Unconstrained to the Constrained Case

In this section we demonstrate that - given that the investor's initial wealth is not too large - the original - but temporarily ignored - constraint on the total number of stocks issued by the firm can approximately be ignored. For the sake of simplicity, we only consider Merton's model. Besides, portfolio problems with vulnerable options are shortly discussed.

[7] See e.g. Harrison (1985) or Musiela/Rutkowski (1999), pp. 211f.

[8] See also Karlin/Taylor (1981), pp. 191ff, who stated requirements without proof.

To analyze the relevance of the constraint, we start with the problem when the investor can actually create a situation where he can "trade" firm value by investing in both stock and bond.

Proposition 5.5 (Relevance of the Bound) *Consider the Merton model and let $x_0 > 0$ be the initial wealth of an investor who maximizes utility from terminal wealth with respect to a power utility function $U(x) = \frac{1}{\gamma}x^\gamma$. Further, assume that*

$$\frac{\lambda}{(1-\gamma)\sigma^2} \leq \frac{v_0}{x_0},$$

i.e. at time $t = 0$ the investor does not plan to buy more securities than issued. Then the probability that his optimal fraction of wealth never exceeds the value of the firm V before T is given by

$$\mathcal{N}\left(\frac{\ln(c) - aT}{|b|\sqrt{T}}\right) - c^{\frac{2a}{b^2}}\mathcal{N}\left(\frac{-\ln(c) - aT}{|b|\sqrt{T}}\right)$$

with

$$a = -\lambda + 0.5\left(\sigma^2 + \tfrac{\lambda^2}{(1-\gamma)\sigma^2}\right), \quad b = \tfrac{\lambda}{(1-\gamma)\sigma} - \sigma, \quad c = \tfrac{\sigma^2 v_0(1-\gamma)}{\lambda x_0}.$$

In the case of a logarithmic utility function $U(x) = \ln(x)$ the result is still valid by taking $\gamma = 0$.

Proof. Since we have

$$V(t) = v_0 \cdot \exp\left((r + \lambda - 0.5\sigma^2)t + \sigma W(t)\right)$$
$$X^*(t) = x_0 \cdot \exp\left((r + 0.5\tfrac{\lambda^2}{\sigma^2})t + \tfrac{\lambda}{\sigma}W(t)\right),$$

in the case of $U(x) = \ln(x)$ the above probability is given by

$$\begin{aligned}
&P\left(\tfrac{\lambda}{\sigma^2}X^*(t) < V(t)\ \forall\, t \in [0,T]\right)\\
&= P\left(\max_{0\leq t\leq T} \tfrac{\lambda}{\sigma^2}\tfrac{x_0}{v_0}\exp\left(0.5[\tfrac{\lambda^2}{\sigma^2} + \sigma^2 - 2\lambda]t + [\tfrac{\lambda}{\sigma} - \sigma]W(t)\right) < 1\right)\\
&= P\Big(\max_{0\leq t\leq T} \underbrace{[-\lambda + 0.5(\sigma^2 + \tfrac{\lambda^2}{\sigma^2})]}_{=a}t + \underbrace{[\tfrac{\lambda}{\sigma} - \sigma]}_{=b}W(t) < \ln(\underbrace{\tfrac{\sigma^2 v_0}{\lambda x_0}}_{=c})\Big)\\
&= P\left(\max_{0\leq t\leq T} at + |b|W(t) < \ln(c)\right)\\
&= P\left(\max_{0\leq t\leq T} \tfrac{a}{|b|}t + W(t) < \tfrac{\ln(c)}{|b|}\right)\\
&= \mathcal{N}\left(\frac{\ln(c) - aT}{|b|\sqrt{T}}\right) - c^{\frac{2a}{b^2}}\mathcal{N}\left(\frac{-\ln(c) - aT}{|b|\sqrt{T}}\right)
\end{aligned}$$

Note that bW and $|b|W$ have the same distributions. The last equation follows from Korn/Korn (2001, p. 168) and is thus a consequence of the reflection principle. The computations for the case $U(x) = \frac{1}{\gamma}x^{\gamma}$ are similar. $\square$

Remarks.

a) It can be shown that for $x_0/v_0 \to 0$, which is the typical situation for a small investor, the above probability approaches one quite fast. Table 5.2 contains numerical values which illustrate this fact. We have taken $\alpha = 0.15$, $r = 0.05$, $\sigma = 0.2$, and $\gamma = 0$. Thus, in the class relevant to a small investor, i.e. $x_0/v_0 \leq 0.01$, the bounds on available numbers of stocks and corporate bonds are virtually irrelevant.

b) If, however, we consider big funds investing in that particular firm, then the results of Table 5.2 indicate equally that the constraints on the total numbers of stocks and corporate bonds cannot be ignored. We shall therefore analyze the constrained problem in the following section.

Table 5.2: Probability of not-touching the barrier for a portfolio problem with stocks and bonds. This table reports probability that an investor who puts his wealth into the money market account and into stocks and bonds of a firm does not plan to buy the entire issues of the firm's equity and debt. The first row indicates different investment horizons T. The following rows report the probabilities for different ratios of initial wealth x_0 to initial firm value v_0. It becomes obvious that for small investors probability is almost one.

Time	Ratio x_0/v_0										
T	0.4	0.35	0.3	0.25	0.2	0.15	0.1	0.05	0.01	0.005	0.001
1	0	0.301	0.612	0.853	0.971	0.998	≈ 1	≈ 1	≈ 1	≈ 1	≈ 1
3	0	0.153	0.338	0.545	0.748	0.906	0.985	≈ 1	≈ 1	≈ 1	≈ 1
5	0	0.107	0.241	0.402	0.586	0.774	0.926	0.995	≈ 1	≈ 1	≈ 1
7	0	0.083	0.188	0.319	0.479	0.662	0.847	0.977	≈ 1	≈ 1	≈ 1
10	0	0.062	0.141	0.243	0.373	0.537	0.732	0.926	0.999	≈ 1	≈ 1

The above proposition is valid only in those cases where the investor allocates his wealth to both stock and bond of the firm. If he is only investing in either stock or bond, then of course the probability increases that the constraints will be binding. However, as both stock and bond prices are non-linear functions of firm value, we cannot expect a closed-form solution for the corresponding probabilities as in the case considered in Proposition 5.5. To illustrate the behavior of these probabilities, we list results obtained via Monte Carlo simulation. As above we have taken $\alpha = 0.15$, $r = 0.05$, $\sigma = 0.2$, $\gamma = 0$, $v_0 = 1000$, and $F = 750$. Besides, it has been assumed that the maturity of the corporate securities occurs one year after the investment horizon.

Table 5.3: Probability of not-touching the barrier for a portfolio problem with corporate bonds. This table reports probability that an investor who puts his wealth into the money market account and into corporate bonds does not plan to buy the entire issue of debt. The first row indicates different investment horizons T. The following rows report the probabilities for different ratios of initial wealth x_0 to initial firm value v_0. It becomes obvious that even for relative small investors probability is far from being one.

Time	Ratio x_0/v_0				
T	0.01	0.005	0.0001	0.00005	0.00001
1	0.151	0.271	0.523	0.617	0.782
3	0.076	0.123	0.226	0.273	0.372
5	0.042	0.066	0.124	0.143	0.207
7	0.022	0.040	0.070	0.090	0.126
10	0.008	0.017	0.032	0.042	0.059

Note that the capital structure of the firm - represented by the face value F of the bond - becomes relevant if the investor allocates his funds to the money market account and to only one of the corporate securities. Since trading only in stocks leads to similar results as before - in fact, every probability in Table 5.3 would be almost one - we omit the corresponding table. In contrast to these results, restricting investment to defaultable bonds heavily increases the probability of touching the barrier. The reason is as follows: If an investor who puts his wealth only into corporate bonds and into the money market account, then his demand for bonds is lower than the supply if[9]

$$\frac{1}{1-\gamma}\frac{\lambda}{B_v\sigma^2} \leq \frac{V}{X}. \tag{5.23}$$

[9] In the following section, we will discuss this inequality in detail.

Omitting discounting a face value of $F = 750$ and an initial firm value of $v_0 = 1000$ results in an equity ratio of 25%, which seems to be a realistic value. But in this case the put inherent in debt is almost valueless and debt is almost riskless. This leads to a delta of the firm's bond which is almost zero. Hence, if an investor plans to take that optimal risky position which is given by Merton's result, he has to buy a large number of bonds. Therefore, the upper bound on the total number of bonds will be violated. This can be seen in inequality (5.23), where the delta of the firm's bond stands in the denominator of the fraction on the left-hand side. By contrast, the delta of the firm's stock is almost one because equity corresponds to a call which is deeply in the money. Hence, investing in stocks leads to a substantially lower probability that the bound is touched.

We end this section with examples of defaultable assets where it is sufficient to consider the unconstrained case. Johnson/Stulz (1987) investigated assets with counterparty risk such as vulnerable options. Their model is similar to Merton's model, as they assume that the value of a vulnerable call at maturity is given by

$$\tilde{C}(T) = \min\{\tilde{V}(T), \max\{\tilde{S}(T) - K, 0\}\},$$

where $\tilde{V}$ denotes the value of the assets of the call writer and $\tilde{S}$ denotes the stock price. Note that in this context the stock price is not a contingent claim on firm value but a primary asset with dynamics

$$d\tilde{S} = \tilde{S}\left[\tilde{\alpha}dt + \tilde{\sigma}dW\right], \quad \tilde{S}(0) = \tilde{s}_0,$$

as in the papers by Black/Scholes (1973) and Merton (1973). Therefore, one has to distinguish between the value of the firm V and the stock price S in Merton's model and the variable $\tilde{V}$ and the stock price $\tilde{S}$ in the model of Johnson/Stulz (1987). Besides, it is important to note that in the present setting there is no link between the assets of the call writer and the number of calls written on the stock $\tilde{S}$. Therefore, the important but inconvenient feature of Merton's model that the total numbers of stocks and bonds are bounded - this would correspond to bounds on the total number of vulnerable calls - is not present in the model by Johnson/Stulz (1987). Consider an investor with power utility function who can put funds into the money market account and vulnerable calls. His optimal fraction invested in vulnerable calls is given by

$$\pi^*_{\tilde{C}}(t) = \frac{1}{1-\gamma}\frac{\tilde{\lambda}}{\tilde{\sigma}^2}\frac{1}{\varepsilon_{\tilde{C}}(t)}.$$

In general, there exists no closed-form solution for the price $\tilde{C}$ of the vulnerable call. Nevertheless, our result remains valid, but we can compute the elasticity $\varepsilon_{\tilde{C}}$ only numerically. For the special case of covered call writing, which means that $\tilde{V}(t) = \rho \cdot \tilde{S}(t)$, $0 < \rho < 1$, Johnson/Stulz (1987) provide a closed-form solution which reads[10]

$$\tilde{C}(0) = C(\tilde{S}(0), K) - (1-\rho) \cdot C(\tilde{S}(0), K/(1-\rho)),$$

where $C(s, K)$ denotes the ordinary Black-Scholes price of a call with strike price K given that the stock price equals s. Hence, we have

$$\frac{\partial \tilde{C}(t)}{\partial \tilde{S}} = \mathcal{N}(d_1'(t)) - (1-\rho) \cdot \mathcal{N}(d_1''(t))$$

with

$$d_1'(t) = \frac{\ln(\frac{\tilde{S}(t)}{K}) + (r + 0.5\tilde{\sigma}^2)(T-t)}{\tilde{\sigma}\sqrt{T-t}},$$

$$d_1''(t) = \frac{\ln(\frac{(1-\rho)\tilde{S}(t)}{K}) + (r + 0.5\tilde{\sigma}^2)(T-t)}{\tilde{\sigma}\sqrt{T-t}}.$$

Using this result, the optimal fraction invested in vulnerable options has the following form

$$\pi_{\tilde{C}}^*(t) = \frac{1}{1-\gamma} \frac{\tilde{\lambda}}{\tilde{\sigma}^2} \frac{\tilde{S}(t)}{\tilde{C}(t) \cdot (\mathcal{N}(d_1'(t)) - (1-\rho) \cdot \mathcal{N}(d_1''(t)))}.$$

Note that this fraction is positive. Clearly, similar results can be calculated in the model by Hull/White (1995), who were able to prove a Black-Scholes-like formula for vulnerable options in a special case of their model.

Another important example of defaultable claims to be considered is that of credit derivatives.

5.4 The Constrained Case

As already announced, we now consider firm value models including all the constraints on the total numbers of stocks and bonds. In this section we discuss

[10] Another closed-form solution is derived for options guaranteed by fixed margins. See Johnson/Stulz (1987) for further details.

Merton's model, the Black-Cox model, and a generalized version of the Briys-de Varenne model.

Denoting the total numbers of stocks and bonds in the investor's portfolio with φ_S and φ_B, respectively, results in the restrictions

$$|\varphi_S(t)| \leq 1 \qquad \text{and} \qquad |\varphi_B(t)| \leq 1.$$

If short sales are prohibited, these constraints become

$$0 \leq \varphi_S(t) \leq 1 \qquad \text{and} \qquad 0 \leq \varphi_B(t) \leq 1.$$

Note that the numbers of stocks and bonds issued by the firm are normalized to unity. Since

$$\varphi_S = \frac{\pi_S \cdot X}{S} \qquad \text{and} \qquad \varphi_B = \frac{\pi_B \cdot X}{B}, \tag{5.24}$$

where X denotes the total wealth of the investor, we obtain

$$|\pi_S| \leq \frac{S}{X} \qquad \text{and} \qquad |\pi_B| \leq \frac{B}{X}$$

or

$$0 \leq \pi_S \leq \frac{S}{X} \qquad \text{and} \qquad 0 \leq \pi_B \leq \frac{B}{X}, \tag{5.25}$$

respectively. Hence, the portfolio process (π_S, π_B) is bounded by a stochastic process which itself depends on the control. Such boundaries do not fall even in the classes treated by Karatzas/Cvitanic (1992). Obviously, these boundaries will be restrictive if the investor's wealth is large enough. Without loss of generality we concentrate on the case when short sales are prohibited.

5.4.1 Merton Model

As before, we consider a portfolio problem in which the investor can put funds into the riskless money market account M, stocks S, and defaultable bonds B. However, we restrict the analysis to the case of an investor maximizing utility from terminal wealth at time T, $T < T_B$, with respect to a logarithmic utility function. To solve the problem, we use the elasticity approach to portfolio optimization. Hence, the wealth equation has the following form:

$$dX(t) = X(t)\Big[(r + \varepsilon(t)\lambda)dt + \varepsilon(t)\sigma dW(t)\Big]. \tag{5.26}$$

Recall that the elasticities of stock and bond are

$$\varepsilon_S = \frac{\partial S}{\partial V}\frac{V}{S} = \mathcal{N}(d_1)\cdot\frac{V}{S}, \qquad \varepsilon_B = \frac{\partial B}{\partial V}\frac{V}{B} = \mathcal{N}(-d_1)\cdot\frac{V}{B}.$$

If the investor can trade in stocks and in bonds, at time $t \in [0,T]$, he can attain any elasticity $\varepsilon(t)$ with

$$\varepsilon(t) = \pi_S(t)\cdot\varepsilon_S(t) + \pi_B(t)\cdot\varepsilon_B(t), \tag{5.27}$$

where the restrictions (5.25) on the portfolio process $\pi = (\pi_S, \pi_B)$ have to be satisfied. Substituting (5.24) into (5.27) leads to

$$\varepsilon(t) = \tfrac{V(t)}{X(t)}\Big(\varphi_S(t)S_v(t) + \varphi_B(t)B_v(t)\Big). \tag{5.28}$$

As S_v, $B_v > 0$, the investor can achieve elasticities $\varepsilon(t)$ which belong to the following interval:

$$[0, \tfrac{V(t)}{X(t)}(S_v + B_v)] = [0, \tfrac{V(t)}{X(t)}]. \tag{5.29}$$

Note that the equality results from the accounting equation $S+B = V$. Hence, we need to solve the following optimization problem

$$\max_{\varepsilon(\cdot)\in\mathcal{A}^*_B(0,x_0)} \mathrm{E}(\ln X^\varepsilon(T)) \tag{5.30}$$

with

$$dX^\varepsilon(t) = X^\varepsilon(t)\Big[(r + \varepsilon(t)\lambda)dt + \varepsilon(t)\sigma dW(t)\Big],$$
$$X^\varepsilon(0) = x_0$$

and $\mathcal{A}^*_B(0,x_0) := \Big\{\varepsilon(\cdot) \in \mathcal{A}(0,x_0) \;:\; X^\varepsilon(t) \geq 0 \text{ and } \varepsilon(t) \in [0,\mathcal{U}(t)] \text{ for } t \in [0,T]\Big\}$, where $\mathcal{U}$ denotes an adapted upper bound for the attainable elasticities and $\mathcal{A}(0,x_0)$ denotes the set of all admissible controls given the initial condition $(0,x_0)$. As shown above, we have $\mathcal{U}(t) = V(t)/X^\varepsilon(t)$ if the investor can trade in both stock and bond. To point out the dependency of X on ε, we used the notation X^ε. In the following, as before, we will mostly omit the superindex. We emphasize that the specific form of the coefficients in the wealth equation guarantees the positivity of $X^\varepsilon(t)$. Besides, we wish to stress that in the present setting the unconstrained optimal elasticity is not always attainable. To emphasize this fact, the elasticity solving the optimization problem (5.30) is said to be the optimal *attainable* elasticity.

The following proposition gives the solution to problem (5.30). Hence, the first step of the elasticity approach is taken.

Proposition 5.6 (Optimal Elasticity in the Merton Model) *Consider the portfolio problem (5.30) and Merton's firm value model. Then the optimal attainable elasticity reads as follows:*

$$\varepsilon^*(t) = \begin{cases} \frac{\lambda}{\sigma^2} & \text{if} \quad \mathcal{U}(t) \geq \frac{\lambda}{\sigma^2}, \\ \mathcal{U}(t) & \text{if} \quad \mathcal{U}(t) < \frac{\lambda}{\sigma^2}. \end{cases} \tag{5.31}$$

Proof. The solution of the wealth equation (5.26) is given by

$$X(t) = x_0 \cdot \exp\Big(\int_0^t r + \varepsilon(s)\lambda - 0.5\varepsilon^2(s)\sigma^2 \, ds + \int_0^t \varepsilon(s)\sigma \, dW(s) \Big).$$

Hence, the investor's expected terminal utility reads

$$\mathrm{E}(\ln X(T)) = \ln x_0 + rT + \mathrm{E}\Big(\int_0^t \varepsilon(s)\lambda - 0.5\varepsilon^2(s)\sigma^2 \, ds \Big).$$

Note that for an admissible control $\varepsilon(\cdot)$ the expectation of the Ito integral vanishes. Completing the square in the above integrand leads to

$$\begin{aligned} \mathrm{E}(\ln X(T)) &= \ln x_0 + rT + 0.5\tfrac{\lambda^2}{\sigma^2}T - 0.5 \cdot \mathrm{E}\Big(\int_0^t \big[\varepsilon(s)\sigma - \tfrac{\lambda}{\sigma}\big]^2 ds \Big) \\ &= \ln x_0 + rT + 0.5\tfrac{\lambda^2}{\sigma^2}T - 0.5\sigma^2 ||\varepsilon - \tfrac{\lambda}{\sigma^2}||^2, \end{aligned}$$

where $||\cdot||$ denotes the norm of the space $\mathcal{L}^2$. Therefore, utility reaches its maximum when the norm is minimal. Using an orthogonal projection argument, the norm becomes minimal if

$$|\varepsilon(t) - \tfrac{\lambda}{\sigma^2}|$$

reaches its minimum at each time instant $t \in [0, T]$. Since $\varepsilon(t) \in [0, \mathcal{U}(t)]$, this leads to (5.31). □

Remarks.

a) We wish to point out that the orthogonal projection argument of the above proof cannot be applied if we assume that the preferences of the investor are governed by a power utility function, since the corresponding integrand is not independent of the wealth process X itself.

b) The above result is not only reasonable, but it can also be applied to more general upper bounds depending on the control ε itself.

By specifying the upper bound $\mathcal{U}$, the optimal portfolio processes can be computed. This corresponds to the second step of the elasticity approach. The respective results are summarized in the following corollary.

Corollary 5.2 (Optimal Portfolios in the Merton Model)
(i) If the investor wishes to put his wealth into stocks and into the riskless money market account, then we have $\mathcal{U}(t) = V(t) \cdot S_v(t)/X(t)$. The optimal fraction invested in the stock is uniquely determined and equals

$$\pi_S^*(t) = \frac{\varepsilon^*(t)}{\varepsilon_S(t)} = \begin{cases} \frac{\lambda}{\sigma^2} \frac{S(t)}{V(t) S_v(t)} & \text{if} \quad \frac{V(t)}{X(t)} S_v(t) \geq \frac{\lambda}{\sigma^2}, \\ \frac{S(t)}{X(t)} & \text{if} \quad \frac{V(t)}{X(t)} S_v(t) < \frac{\lambda}{\sigma^2}, \end{cases} \tag{5.32}$$

where $S_v = N(d_1)$.

(ii) If the investor wishes to put his wealth into defaultable bonds and into the riskless money market account, then we have $\mathcal{U}(t) = V(t) \cdot B_v(t)/X(t)$. The optimal fraction invested in the bond is uniquely determined and equals

$$\pi_B^*(t) = \frac{\varepsilon^*(t)}{\varepsilon_B(t)} = \begin{cases} \frac{\lambda}{\sigma^2} \frac{B(t)}{V(t) B_v(t)} & \text{if} \quad \frac{V(t)}{X(t)} B_v(t) \geq \frac{\lambda}{\sigma^2}, \\ \frac{B(t)}{X(t)} & \text{if} \quad \frac{V(t)}{X(t)} B_v(t) < \frac{\lambda}{\sigma^2}, \end{cases} \tag{5.33}$$

where $B_v = N(-d_1)$.

(iii) If the investor wishes to allocate his wealth to stocks, to defaultable bonds, and to the riskless money market account, then we have $\mathcal{U}(t) = V(t)/X(t)$. The optimal portfolio process is not uniquely determined, but every combination of stock and bond is optimal if it leads to the optimal elasticity.

Proof. (i) Since the investor puts his funds into stocks and the money market account, the attainable elasticities are given by

$$\varepsilon(t) = \frac{V(t)}{X(t)} \varphi_S(t) S_v(t), \quad 0 \leq \varphi_S \leq 1. \tag{5.34}$$

Therefore, we get $\mathcal{U}(t) = V(t) \cdot S_v(t)/X(t)$. Substituting the definition of φ_S into (5.34) leads to (5.32).

As equation (5.34) can uniquely be solved for φ_S, the equation (5.32) is proved.

(ii) The proof of (ii) is similar to the proof of (ii).

(iii) The upper bound follows from (5.29). Since there is one degree of freedom for the solution of equation (5.28), the portfolio process is not uniquely determined. □

Remark.
The situation described in (iii) has an interesting geometric interpretation. The optimal combinations of stock and bond lie on the straight line $\mathcal{G}$ given by

$$\pi_S = \tfrac{1}{\varepsilon_S}\Big(\tfrac{\lambda}{\sigma^2} - \pi_B \varepsilon_B\Big),$$

whereas the range of all attainable strategies (π_S, π_B) corresponds to the rectangle

$$\mathcal{R} := [0, \tfrac{S}{X}] \times [0, \tfrac{B}{X}].$$

Hence, the unconstrained optimal elasticity λ/σ^2 is attainable if the straight line and the rectangle have at least one point in common, i.e. $\mathcal{G} \cap \mathcal{R} \neq \emptyset$.

As mentioned in the previous section, the ratio v_0/x_0 between initial firm value and the investor's initial wealth is crucial for the relevance of the restriction. An investor who has extensive funds to invest - i.e. his initial wealth is large - will not be able to buy a position which perfectly tracks the unconstrained optimal elasticity. Hence, a *tracking error* occurs. Clearly, this problem will be less severe if firm value increases. Nevertheless, such restrictions cannot be neglected in the case of large investment funds.

5.4.2 Black-Cox Model

In this subsection we make the same basic assumptions as in Subsection 5.4.1. In particular, we consider an investor with logarithmic utility function. Besides, it is assumed for the sake of convenience that risk-neutral valuation formulae for the securities issued by the firm are valid. As in Subsection 5.2.3 we assume that the only attainable elasticity after default is zero. Thus the portfolio problem of an investor reads

$$\max_{\varepsilon(\cdot) \in \mathcal{A}'_B(0,x_0)} \mathrm{E}(\ln X^{\varepsilon}(T))\} \tag{5.35}$$

with

$$\begin{aligned} dX^{\varepsilon}(t) &= X^{\varepsilon}(t)\Big[(r + \varepsilon(t)\lambda)dt + \varepsilon(t)\sigma dW(t)\Big], \\ X^{\varepsilon}(0) &= x_0, \end{aligned}$$

and

$$\mathcal{A}'_B(0,x_0) := \Big\{\varepsilon(\cdot) \in \mathcal{A}^*(0,x_0) \,:\, \varepsilon(t) = 0 \,\forall\, t \in [\tau, T]\Big\}.$$

Note that the elements of $\mathcal{A}'_B(0,x_0)$ are processes which are killed at the stopping time τ. Hence, $\mathcal{A}'_B(0,x_0) \subset \mathcal{A}^*_B(0,x_0)$, which implies that all elements of $\mathcal{A}'_B(0,x_0)$ are admissible controls. We emphasize that the partial derivatives of both stock and bond with respect to firm value are positive. Hence, only positive elasticities are attainable. The solution of this problem is given in the following proposition.

Proposition 5.7 (Optimal Elasticity in the Black-Cox Model) *If we consider the portfolio problem (5.35) for the Black-Cox model, then the optimal attainable elasticity equals*

$$\varepsilon^*(t) = \begin{cases} \frac{\lambda}{\sigma^2} & \text{if} \quad \tau > t \text{ and } \mathcal{U}(t) \geq \frac{\lambda}{\sigma^2}, \\ \mathcal{U}(t) & \text{if} \quad \tau > t \text{ and } \mathcal{U}(t) < \frac{\lambda}{\sigma^2}, \\ 0 & \text{if} \quad \tau \leq t. \end{cases}$$

Proof. Similar arguments apply as in the proof of Proposition 5.6. □

Since the investor has logarithmic utility, we can apply a pointwise maximization argument. Hence, for the optimality of elasticity up to default it is irrelevant if or when default occurs. Consequently, the portfolio remains optimal regardless of whether the firm is liquidated after default, is taken over, or is reorganized and new securities are issued etc. For this reason it is sufficient to consider the optimization problem (5.35), where the attainable elasticity ε is killed after default. Without this independence one needs to take into account whether there are any tradable securities of the firm, once default has occurred.

Applying Proposition 5.7, we can determine the optimal portfolio fractions.

Corollary 5.3 (Optimal Portfolios in the Black-Cox Model)
(i) If the investor wishes to put his wealth into stocks and into the riskless money market account, then we obtain $\mathcal{U}(t) = V(t) \cdot S_v(t)/X(t)$, $0 \leq t < \tau$. The optimal fraction invested in stocks is uniquely determined and equals

$$\pi^*_S(t) = \frac{\varepsilon^*(t)}{\varepsilon_S(t)} = \begin{cases} \frac{\lambda}{\sigma^2}\frac{S(t)}{V(t)S_v(t)} & \text{if} \quad \tau > t \text{ and } \frac{V(t)}{X(t)}S_v(t) \geq \frac{\lambda}{\sigma^2}, \\ \frac{S(t)}{X(t)} & \text{if} \quad \tau > t \text{ and } \frac{V(t)}{X(t)}S_v(t) < \frac{\lambda}{\sigma^2}, \\ 0 & \text{if} \quad \tau \leq t, \end{cases}$$

where $S_v(t) = \mathcal{N}(z_3(t)) + y^{2\theta}(t)\left[\frac{2\theta-2}{V(t)y^2(t)}C(t) + \mathcal{N}(z_4(t))\right]$.

(ii) If the investor wishes to put his wealth into defaultable bonds and into the riskless money market account, then we obtain $\mathcal{U}(t) = V(t) \cdot B_v(t)/X(t)$, $0 \le t < \tau$. *The optimal fraction invested in bonds is uniquely determined and equals*

$$\pi_B^*(t) = \frac{\varepsilon^*(t)}{\varepsilon_B(t)} = \begin{cases} \frac{\lambda}{\sigma^2}\frac{B(t)}{V(t)B_v(t)} & \text{if} \quad \tau > t \text{ and } \frac{V(t)}{X(t)}B_v(t) \ge \frac{\lambda}{\sigma^2}, \\ \frac{B(t)}{X(t)} & \text{if} \quad \tau > t \text{ and } \frac{V(t)}{X(t)}B_v(t) < \frac{\lambda}{\sigma^2}, \\ 0 & \text{if} \quad \tau \le t, \end{cases}$$

where $B_v(t) = \mathcal{N}(-z_3(t)) - y^{2\theta}(t)\left[\frac{2\theta-2}{V(t)y^2(t)}C(t) + \mathcal{N}(z_4(t))\right]$.

(iii) If the investor wishes to allocate his wealth to defaultable bonds, to stocks, and to the riskless money market account, then $\mathcal{U}(t) = V(t)/X(t)$, $0 \le t < \tau$. *The optimal portfolio process is not uniquely determined. Every combination of stock and bond is optimal if it leads to the optimal elasticity.*

Remark.

To motivate the elasticity approach, in Chapter 3 we applied the Black-Scholes partial differential equation. Note that barrier derivatives meet this equation up to the stopping time τ. For this reason the elasticity approach is still applicable to the Black-Cox model.

As mentioned in the introduction, Black/Cox (1976) also considered the case of subordinated bonds (syn. junior bonds). Since the value of such corporate bonds can be expressed as the difference between the value of two senior bonds, junior bonds also correspond to contingent claims on firm value. Therefore, portfolio problems with junior bonds can be solved by a similar approach.

5.4.3 Generalized Briys-de Varenne Model

One shortcoming of the Black-Cox model results from their assumption that interest rates are deterministic, although corporate bonds may be significantly influenced by interest rate risk.[11] To overcome this drawback Briys/de

[11] See e.g. Jones/Mason/Rosenfeld (1984), Kim/Ramaswamy/Sundaresan (1993), or Longstaff/Schwartz (1995).

Varenne (1997) assumed to short rate to be stochastic, following the dynamics of the extended Vasicek model. Clearly, this model is applied to get closed-form solutions for the values of the firm's securities. Since we do not need this assumption, we simply assume that the dynamics of the short rate are governed by

$$dr(t) = a(t)dt + b(t)dW_r(t), \quad r(0) = r_0 > 0, \tag{5.36}$$

where, with a slight abuse of notation, the drift $a(t) = a(t, r(t))$ and the volatility $b(t) = b(t, r(t))$ are assumed to be measurable. Besides, we assume that $\int_0^t |a(s)| \, ds < \infty$ and

$$\mathrm{E}\left(\int_0^T b^4(s) \, ds \right) < \infty \tag{5.37}$$

and that the SDE (5.36) has a unique solution.[12] Hence, the models by Vasicek (1977), Dothan (1978), Cox/Ingersoll/Ross (1985), Ho/Lee (1986), and Black/Karasinski (1991) are special cases of (5.36). Additionally, let $b \neq 0$ almost surely.

The dynamics of firm value are modeled by

$$dV(t) = V(t)\Big[\Big(r(t) + \lambda_V(t)\Big)dt + \sigma_V(t)dW_V(t) + \sigma_r(t)dW_r(t)\Big],$$

where the processes $\lambda_V(t) = \lambda_V(t, V(t), r(t))$, $\sigma_V(t) = \sigma_V(t, V(t), r(t))$, and $\sigma_r(t) = \sigma_r(t, V(t), r(t))$ are assumed to be measurable. Besides, we assume that $\int_0^t |\lambda_V(s)| \, ds < \infty$ and

$$\mathrm{E}\left(\int_0^T \sigma_V^4(s) + \sigma_r^4(s) \, ds \right) < \infty \tag{5.38}$$

and that the SDE (5.36) has a unique solution. In addition, let σ_V and σ_r be bounded away from zero.

Following the lines of Black/Cox (1976), Briys/de Varenne (1997) allowed for immediate default if a lower (discounted) boundary

$$L(t) = k \cdot F \cdot P(t, T_B)$$

is reached, where $P(t, T_B)$ denotes the price of a riskless bond with maturity T_B at time t. If default does not occur prior to T_B, it is assumed that the

[12] See Proposition 1.1 and 1.2 as well as Theorem 1.1.

corporate bond has the same terminal value as in the Merton model or the Black-Cox model. Note that for deterministic interest rates the above boundary is equal to $kFe^{-r(T_B-t)}$, which corresponds to a special case of the default boundary in the Black-Cox model. Hence, the value of the corporate bond can now be expressed as the value of a portfolio consisting of a down-and-out put, a down-and-out bond, and a down-and-in bond. More precisely, the value of the defaultable bond at time T_B is given by[13]

$$B(T_B, V(T_B)) = F \cdot P_{DO}(T_B) - \max\{F - V(T_B), 0\}_{DO} + k \cdot F \cdot P_{DI}(T_B),$$

where P_{DI} denotes the down-and-in bond with knock-in barrier L. Recall that in the case of a discounted barrier the values of cash-at-hit options and down-and-in bonds coincide. Besides, we have $P_{DO}(T_B) = 1_{\{\tau > T_B\}}$ and $P_{DI}(T_B) = 1_{\{\tau \leq T_B\}}$, where 1_E denotes the indicator function for E and $\tau = \inf\{t \geq 0 : V(t) = L(t)\}$. The stock price at maturity T_B of the corporate bond reads

$$S(T_B, V(T_B)) = \max\{V(T_B) - F, 0\}_{DO},$$

i.e. equity corresponds to a down-and-out call with discounted barrier. Since both stock and bond are contingent claims depending on firm value and short rate, we proceed by extending the elasticity approach to stochastic interest rates. In the previous sections we have seen that the elasticity of contingent claims with respect to firm value plays an important role. In this section a second Brownian motion has been introduced, namely the one which governs the interest rate risk. As a consequence, the sensitivity of a claim with respect to interest rate risk has become relevant. This sensitivity is said to be the duration of the claim and is defined by $D_C = C_r/C$.[14]

Keeping this in mind, we consider a contingent claim $C(t) = C(t, V(t), r(t))$. Assuming sufficient differentiability, an application of Ito's formula leads to

[13] Briys/de Varenne (1997) also allowed for deviations of the absolute priority rule. We do not consider this generalization here because it does not fundamentally alter our results.

[14] In the context of stochastic interest rates duration was introduced by Cox/Ingersoll/Ross (1979) using their interest rate model. They define duration as $G^{-1}(-D_C)$, where G is a model dependent function. This is due to the fact that duration is normally measured in units of time. Only if interest rates are deterministic or the time-continuous Ho-Lee model is used do both definitions coincide. Since in our context only the sensitivity C_r/C becomes relevant, we use the word "duration" for this sensitivity in accordance with the deterministic case.

$$\begin{aligned} dC &= C_t dt + C_v dV + C_r dr + 0.5 C_{vv} d < V > +0.5 C_{rr} d < r > \\ &\quad + C_{vr} d < V, r > \\ &= \Big(C_t + C_v V(r + \lambda_V) + aC_r + 0.5 C_{vv} V^2(\sigma_V^2 + \sigma_r^2) + 0.5 C_{rr} b^2 \\ &\quad + b\sigma_r V C_{vr}\Big) dt + C_v V \sigma_V dW_V + (C_v V \sigma_r + C_r b) dW_r. \end{aligned} \tag{5.39}$$

Note that the claim price satisfies the following generalized Black-Scholes partial differential equation:[15]

$$\begin{aligned} &C_t + \Big(r + \lambda_V - \sigma_V \zeta_V - \sigma_r \zeta_r\Big) v C_v + \Big(a - \zeta_r b\Big) C_r \\ &+0.5\Big((\sigma_V^2 + \sigma_r^2) v^2 C_{vv} + 2b\sigma_r v C_{vr} + b^2 C_{rr}\Big) - rC = 0, \end{aligned}$$

where $\zeta_V := \lambda_V / \sigma_V$ denotes the market price of risk of firm value and $\zeta_r(t) = \zeta_r(t, r(t))$ stands for the market price of risk of the market for riskless bonds. We assume that the latter is measurable and integrable. The market price ζ_r remains unspecified as long as one does not assume that an interest rate sensitive claim, e.g. a riskless bond, is traded. Defining the aggregate excess return $\lambda := \sigma_V \zeta_V + \sigma_r \zeta_r$ and substituting the generalized Black-Scholes partial differential equation into (5.39) gives

$$dC = C\Big[\big(r + \lambda \varepsilon_C + b\zeta_r D_C\big) dt + \sigma_V \varepsilon_C dW_V + \big(\sigma_r \varepsilon_C + bD_C\big) dW_r\Big],$$

where $\varepsilon_C := C_v V / C$ denotes the elasticity of the claim and $D_C := C_r / C$ the duration of the claim. Let π_C denote the fraction invested in the contingent claim, then the number of claims in the investor's portfolio is given by $\varphi_C = \pi_C \cdot C / X$. If an investor can only divide his wealth between the riskless money market account and the claim, the amount of money invested in the account equals $\varphi_M = (1 - \pi_C) \cdot M / X$. Therefore, we arrive at the wealth equation

$$\begin{aligned} dX &= \varphi_M dM + \varphi_C dC \\ &= (1 - \pi_C) X r dt \\ &\quad + \pi_C X\Big[\big(r + \lambda \varepsilon_C + b\zeta_r D_C\big) dt + \sigma_V \varepsilon_C dW_V + \big(\sigma_r \varepsilon_C + bD_C\big) dW_r\Big], \\ &= X\Big[\big(r + \lambda \varepsilon + b\zeta_r D\big) dt + \sigma_V \varepsilon dW_V + \big(\sigma_r \varepsilon + bD\big) dW_r,\Big] \end{aligned}$$

where ε and D denote the static elasticity and duration of the portfolio, i.e.

$$\varepsilon = \pi_C \cdot \varepsilon_C + (1 - \pi_C) \cdot \varepsilon_M \quad \text{and} \quad D = \pi_C \cdot D_C + (1 - \pi_C) \cdot D_M. \tag{5.40}$$

[15] See Merton (1973) or Duffie (1992) for a textbook reference.

Recall that $\varepsilon_M = D_M = 0$. This procedure can be generalized to an arbitrary number of claims $C(t) = C(t, V(t), r(t))$. Hence, the elasticity approach is applicable to the case of stochastic interest rates. The relevant controls are elasticity and duration.

Given an investor with logarithmic utility function, the optimal portfolio problem reads

$$\max_{(\varepsilon(\cdot),D(\cdot))\in\mathcal{A}'_B(0,x_0)} \mathrm{E}\Big(\ln X^{\varepsilon,D}(T)\Big) \tag{5.41}$$

with

$$\begin{aligned} dX^{\varepsilon,D} = X^{\varepsilon,D}\Big[\Big(r + (\sigma_V\zeta_V + \sigma_r\zeta_r)\varepsilon + \zeta_r bD\Big)dt \\ +\sigma_V\varepsilon dW_V + (\sigma_r\varepsilon + bD)dW_r\Big], \qquad X^{\varepsilon,D}(0) = x_0 \end{aligned}$$

and

$$\begin{aligned} \mathcal{A}'_B(0,x_0) := \Big\{(\varepsilon(\cdot), D(\cdot)) \in \mathcal{A}(0,x_0) \,:\, X^{\varepsilon,D}(t) \geq 0,\, \varepsilon(t) \in [\mathcal{L}_\varepsilon(t), \mathcal{U}_\varepsilon(t)], \\ D(t) \in [\mathcal{L}_D(t), \mathcal{U}_D(t)]\,\forall\, t \in [0,\tau),\, \varepsilon(t) = D(t) = 0\,\forall\, t \in [\tau, T]\Big\}, \end{aligned}$$

where $\mathcal{U}_\varepsilon$ ($\mathcal{U}_D$) denotes an adapted upper bound on the attainable elasticities (durations) and $\mathcal{L}_\varepsilon$, $\mathcal{L}_D$ the corresponding adapted lower bounds. We assume that $\mathcal{L}_\varepsilon < \mathcal{U}_\varepsilon$ and $\mathcal{L}_D < \mathcal{U}_D$. As the dynamics of firm value and short rate are not further specified, we cannot exclude situations where elasticity is negative. By analogy, duration can be positive or negative. Note that in a complete market we would have $\mathcal{U}_\varepsilon = \mathcal{U}_D = +\infty$ and $\mathcal{L}_\varepsilon = \mathcal{L}_D = -\infty$.

Since in the present portfolio problem there are two sources of risk, which are modeled by a two-dimensional Brownian motion, an investor needs generally at least two different risky securities to attain a given combination of elasticity and duration. Nevertheless, two securities might not be sufficient to replicate such a combination. This is due to the restrictions on the attainable elasticities and durations. If an investor can only divide his wealth between the money market account and a corporate bond, he will not in general be able to replicate a given combination of elasticity and duration. In this case, by (5.40), the set $\mathcal{A}'_B(0,x_0)$ has to be further restricted via the condition

$$D = \varepsilon \cdot D_C/\varepsilon_C. \tag{5.42}$$

Consequently, the investor can only choose the portfolio elasticity ε. The duration of his portfolio is then given by condition (5.42). Equivalently, he can decide for a duration, but then the elasticity of his portfolio is automatically fixed. In the present setting, this missing degree of freedom is a characteristic for each portfolio problem with only one risky investment opportunity. Therefore, it will be beneficial to the investor if his investment opportunities include tradable stocks or subordinated bonds. However, if we assume that a firm issues senior and junior bonds, it has to be taken into account that in this case we have $V = S + B^{junior} + B^{senior}$.

We are now in a position to solve the portfolio problems.

Proposition 5.8 (Optimal Sensitivities in the Briys-de V. model) *Consider the portfolio problem (5.41) and the Briys-de Varenne model with generalized stochastic short rate (5.36). Let* $\varepsilon_{uc} := \lambda_V/\sigma_V^2$ *and* $D_{uc} := (\zeta_r - \zeta_V\sigma_r/\sigma_V)/b$ *be the unconstrained optimal elasticity and duration.*

(i) If the investor can invest in at least two corporate securities, such as stocks and bonds, the optimal elasticity and the optimal duration prior to default can be represented by

$$\varepsilon^*|_{[0,\tau]} = \varepsilon_{uc} + \frac{b\cdot(-\chi_1+\chi_2)+\sigma_r\cdot(\chi_3-\chi_4)}{b\sigma_V^2}, \tag{5.43}$$

$$D^*|_{[0,\tau]} = D_{uc} + \frac{b\sigma_r\cdot(\chi_1-\chi_2)+(\sigma_V^2+\sigma_r^2)\cdot(-\chi_3+\chi_4)}{b^2\sigma_V^2}, \tag{5.44}$$

where $\chi_1, \chi_2, \chi_3, \chi_4 \geq 0$ *denote Lagrangian multipliers corresponding to the constraints* $\mathcal{U}_\varepsilon - \varepsilon \geq 0$, $\varepsilon - \mathcal{L}_\varepsilon \geq 0$, $\mathcal{U}_D - D \geq 0$, $D - \mathcal{L}_D \geq 0$. *The precise results for all possible combinations of binding constraints are summarized in Table 5.4.*

(ii) If the investor can only invest in one particular corporate security with elasticity ε_C *and duration* D_C*, then the optimal elasticity prior to default, i.e.* $t \leq \tau$*, is given by*

$$\tilde{\varepsilon}(t) = \begin{cases} \mathcal{L}_\varepsilon(t) & \text{if } \mathcal{L}_\varepsilon(t) > \frac{\lambda(t)+\zeta_r(t)b(t)R(t)}{\sigma_V^2(t)+(\sigma_r(t)+b(t)R(t))^2}, \\ \frac{\lambda(t)+\zeta_r(t)b(t)R(t)}{\sigma_V^2(t)+(\sigma_r(t)+b(t)R(t))^2} & \text{if } \mathcal{U}_\varepsilon(t) \geq \frac{\lambda(t)+\zeta_r(t)b(t)R(t)}{\sigma_V^2(t)+(\sigma_r(t)+b(t)R(t))^2} \geq \mathcal{L}_\varepsilon(t), \\ \mathcal{U}_\varepsilon(t) & \text{if } \mathcal{U}_\varepsilon(t) < \frac{\lambda(t)+\zeta_r(t)b(t)R(t)}{\sigma_V^2(t)+(\sigma_r(t)+b(t)R(t))^2}, \end{cases}$$

where $R := D_C/\varepsilon_C$*. The optimal duration is then given by (5.42).*

Table 5.4: Optimal elasticity and duration in the Briys-de Varenne model. This table reports the optimal elasticity and duration of a portfolio problem where an investor can put his wealth into at least two different corporate securities. Since there are upper and lower bounds on the attainable elasticities and durations, nine cases need to be distinguished. These cases are characterized via the Lagrangian multipliers. For notational convenience we omit the dependencies on time t.

case	Lagrange multipliers				optimal elasticity	optimal duration
no.	χ_1	χ_2	χ_3	χ_4	ε^*	D^*
1	$=0$	$=0$	$=0$	$=0$	λ_V/σ_V^2	$(\zeta_r - \sigma_r\lambda_V/\sigma_V^2)/b$
2	>0	$=0$	$=0$	$=0$	$\mathcal{U}_\varepsilon$	$(\zeta_r - \sigma_r\mathcal{U}_\varepsilon)/b$
3	>0	$=0$	>0	$=0$	$\mathcal{U}_\varepsilon$	$\mathcal{U}_D$
4	>0	$=0$	$=0$	>0	$\mathcal{U}_\varepsilon$	$\mathcal{L}_D$
5	$=0$	>0	$=0$	$=0$	$\mathcal{L}_\varepsilon$	$(\zeta_r - \sigma_r\mathcal{L}_\varepsilon)/b$
6	$=0$	>0	>0	$=0$	$\mathcal{L}_\varepsilon$	$\mathcal{U}_D$
7	$=0$	>0	$=0$	>0	$\mathcal{L}_\varepsilon$	$\mathcal{L}_D$
8	$=0$	$=0$	>0	$=0$	$(\lambda - \sigma_r b\,\mathcal{U}_D)/(\sigma_V^2 + \sigma_r^2)$	$\mathcal{U}_D$
9	$=0$	$=0$	$=0$	>0	$(\lambda - \sigma_r b\,\mathcal{L}_D)/(\sigma_V^2 + \sigma_r^2)$	$\mathcal{L}_D$

Proof. (i) For an admissible control (ε, D) the solution of the wealth equation reads

$$X(T) = x_0 \exp\left(\int_0^T r(s)\,ds\right) \cdot \exp\left(\int_0^T \lambda(s)\varepsilon(s) + \zeta_r(s)b(s)D(s) -0.5\sigma_V^2(s)\varepsilon^2(s) - 0.5\Big(\sigma_r(s)\varepsilon(s) + b(s)D(s)\Big)^2 ds\right)$$
$$\exp\left(\int_0^T \sigma_V(s)\varepsilon(s)\,dW_V(s) + \int_0^T \sigma_r(s)\varepsilon(s) + b(s)D(s)\,dW_r(s)\right)$$

By (5.37) and (5.38), we have for an admissible control (ε, D)

$$\mathrm{E}\left(\int_0^T \Big(\sigma_V(s)\varepsilon(s)\Big)^2 ds\right) \le \mathrm{E}\left(\int_0^T \sigma_V^4(s) + \varepsilon^4(s)\,ds\right) < \infty$$

and therefore $\mathrm{E}\left(\int_0^T \sigma_V(s)\varepsilon(s)\,dW_V(s)\right) = 0$. By analogy, the expected values of the other Ito integrals are zero. Hence, taking expectations leads to

$$
\begin{aligned}
\mathrm{E}(\ln X(T)) = \ln(x_0) + \mathrm{E}\Big(\int_0^T r(s)\,ds\Big) + \mathrm{E}\Big(\int_0^T \lambda(s)\varepsilon(s) + \zeta_r(s)b(s)D(s) \\
-0.5\sigma_V^2(s)\varepsilon^2(s) - 0.5\big(\sigma_r(s)\varepsilon(s) + b(s)D(s)\big)^2\,ds\Big), \qquad (5.45)
\end{aligned}
$$

Obviously, if the integrand of the second integral is pathwise maximized over $\varepsilon(t)$ and $D(t)$, we obtain optimal elasticity and duration. Hence, at time $t < \tau$ we face a constrained optimization problem with concave objective function[16]

$$
\begin{aligned}
f(\varepsilon, D) &:= \lambda\varepsilon + \zeta_r bD - 0.5\sigma_V^2\varepsilon^2 - 0.5\big(\sigma_r\varepsilon + bD\big)^2 \\
&= 0.5(\zeta_V^2 + \zeta_r^2) - 0.5(\zeta_V - \sigma_V\varepsilon)^2 - 0.5(\zeta_r - \sigma_r\varepsilon - bD)^2
\end{aligned}
$$

under the constraints $\mathcal{U}_\varepsilon - \varepsilon \geq 0$, $\varepsilon - \mathcal{L}_\varepsilon \geq 0$, $\mathcal{U}_D - D \geq 0$, $D - \mathcal{L}_D \geq 0$. As a consequence, optimal elasticity and duration meet the following Kuhn-Tucker conditions (KTC)

$$
\begin{aligned}
\lambda - (\sigma_V^2 + \sigma_r^2)\varepsilon - \sigma_r bD - \chi_1 + \chi_2 &= 0, \\
b\zeta_r - \sigma_r b\varepsilon - b^2 D - \chi_3 + \chi_4 &= 0, \\
\chi_1(\mathcal{U}_\varepsilon - \varepsilon) &= 0, \\
\chi_2(\varepsilon - \mathcal{L}_\varepsilon) &= 0, \\
\chi_3(\mathcal{U}_D - D) &= 0, \\
\chi_4(D - \mathcal{L}_D) &= 0
\end{aligned}
$$

with Lagrangian multipliers $\chi_1, \chi_2, \chi_3, \chi_4 \geq 0$. Notice that in our problem the KTCs are also sufficient for optimality. Assume first that $\chi_i = 0$, $i = 1, 2, 3, 4$, (case 1). Then from the first and second KTC we get $\varepsilon^* = \lambda_V/\sigma_V^2$ and $D^* = (\zeta_r - \sigma_r\lambda_V/\sigma_V^2)/b$. If $\chi_1 > 0$, we conclude $\varepsilon^* = \mathcal{U}_\varepsilon$ and $\chi_2 = 0$. Assume further that $\chi_3 = 0$ and $\chi_4 = 0$ (case 2). Solving the second KTC for D leads to $D^* = (\zeta_r - \sigma_r\mathcal{U}_\varepsilon)/b$. The first KTC can be used to check that χ_1 is indeed strict positive. Assuming $\chi_1 > 0$ and $\chi_3 > 0$ leads to $\varepsilon^* = \mathcal{U}_\varepsilon$, $D^* = \mathcal{U}_D$ and $\chi_2 = 0$, $\chi_4 = 0$ (case 3). Using the first and the second KTC one can check that χ_1, χ_3 are indeed strictly positive. The other cases summarized in Table 5.4 can be treated similarly. Besides, the representations (5.43) of ε^* and D^* can be calculated by solving the first and second KTC for elasticity and duration.

(ii) Substituting (5.42) into the function f of (i) leads to an optimization problem over $\varepsilon(t)$. Solving this problem leads to the optimal elasticity $\tilde{\varepsilon}$. □

[16] For notational convenience we omit the dependencies on time t.

Remarks.

a) Note that the unconstrained optimal elasticity ε_{uc} does not depend on traded assets. Hence, the elasticity ε^* only depends on traded assets if one of the bounds is touched, whereas the elasticity $\tilde{\varepsilon}$ is never independent of traded assets.

b) Assuming logarithmic utility leads to separation of the accumulation factor $\int_0^T r(s)\,ds$. This becomes obvious in equation (5.45). Therefore, we can avoid specifying the term structure model further.

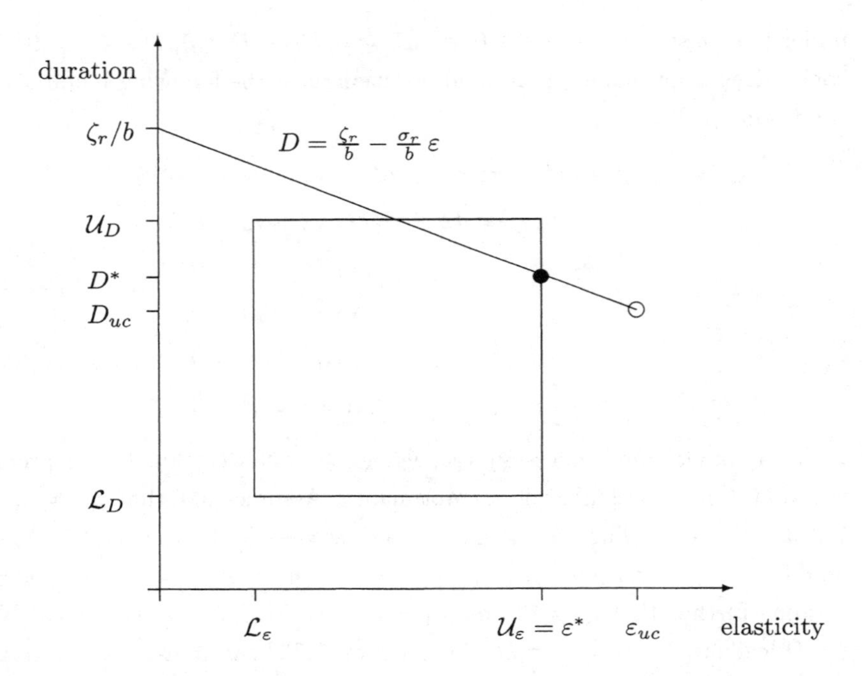

Figure 5.1: Unconstrained optimal elasticity not attainable. This figure illustrates the situation of an investor who is not able to attain the unconstrained optimal elasticity ε_{uc}, but the bounds on duration are not binding. The square restricts the set of attainable combinations of elasticity and duration. His optimal strategy is given by the right-hand intersection of square and straight line $D = \zeta_r/b - \varepsilon\sigma_r/b$ because the bounds on duration are not active.

Of course, the formulation of Proposition 5.8 is not the one which is suitable for practical purposes as Lagrangian multipliers are not observable input parameters. It is always possible to determine from the observable variables in which case of Proposition 5.8 we are. However, a detailed formulation would lead to a large number of cases and subcases which does not allow for a compact presentation. To demonstrate at least one such situation in an explicit form we concentrate on cases 2 and 3 of Proposition 5.8. The other cases mentioned in Table 5.4 can be interpreted in a similar way.

Assume that all parameters are positive. If the investor is forced to take less firm value risk than he wishes to take, i.e. $\varepsilon_{uc} > \mathcal{U}_\varepsilon$, and the constraints on duration are not binding (case 2), his optimal duration D^* equals $(\zeta_r - \sigma_r \mathcal{U}_\varepsilon)/b$, i.e. $D^* > D_{uc}$. Hence, he tries to compensate for the constraint on elasticity by increasing his interest rate exposure. More generally, given the optimal elasticity ε^* his optimal duration reads $(\zeta_r - \sigma_r \varepsilon^*)/b$, i.e. a forced decrease of elasticity leads to an increase of optimal duration. But this adjustment of duration is only possible if the upper bound $\mathcal{U}_D$ is not reached. Otherwise, the corner solution $\varepsilon^* = \mathcal{U}_\varepsilon$ and $D^* = \mathcal{U}_D$ is optimal (case 3). In Figures 5.1 and 5.2 we have illustrated the two cases. Note that the maximal interest rate exposure equals ζ_r/b, which can be interpreted as risk premium of the money market.[17] An increase of this premium makes investment in the money market more attractive. Therefore, the investor is willing to take more interest rate exposure. Moreover, given $\varepsilon_{uc} > \mathcal{U}_\varepsilon$ the excess investment in the money market $(\zeta_r - \sigma_r \mathcal{U}_\varepsilon)/b - D_{uc}$ is greater than in the case involving a lower risk premium. Clearly, these arguments can be formalized by applying the KTC.

We wish to stress that in Figure 5.1 and 5.2 we have assumed that the unconstrained optimum $(\varepsilon_{uc}, D_{uc})$ remains the same, although ζ_r/b is increased in Figure 5.2. Since the straight line describing the substitution effect is given by

$$D = \frac{\zeta_r}{b} - \frac{\sigma_r}{b}\varepsilon,$$

it becomes obvious that the absolute value of the slope $-\sigma_r/b$ has to be increased, too. This can only occur if both ζ_r and σ_r are increased because a one-sided decrease of b would contradict the assumption that the unconstrained optimum remains the same. Consequently, the increase of ζ_r provides for the excess investment in the money market, whereas the increase of σ_r is not relevant to the investor.

---

[17] In this context we use "money market" in a rather loose way and mean a market which is only affected by interest rate risk.

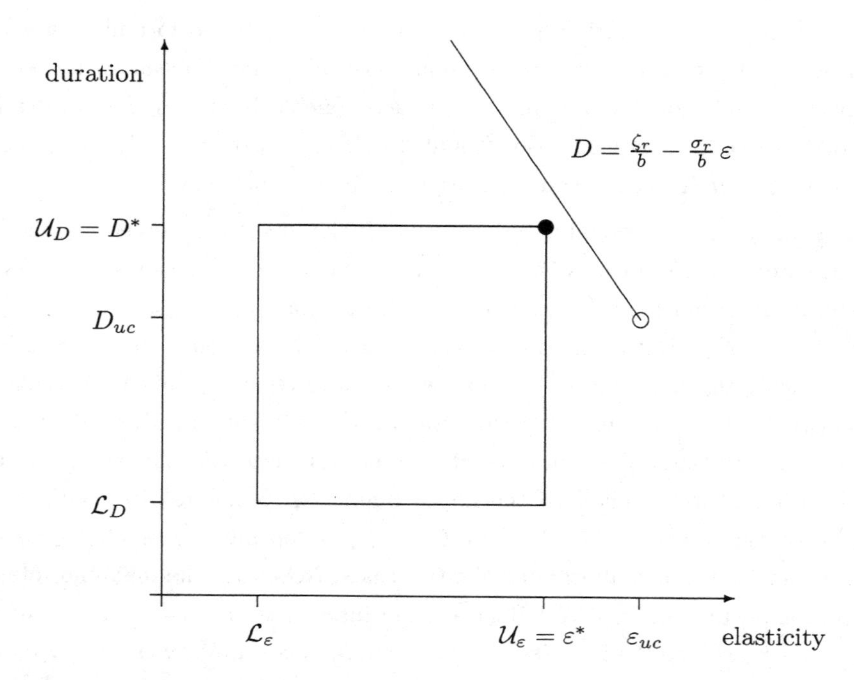

Figure 5.2: Unconstrained optimal elasticity and duration not attainable. This figure illustrates the situation of an investor who is neither able to attain the unconstrained optimal elasticity ε_{uc} nor the unconstrained optimal duration D_{uc}. This is due to the fact that both the upper bound on elasticity and the upper bound on duration are violated. The square restricts the set of attainable combinations of elasticity and duration. Since the straight line $D = \zeta_r/b - \varepsilon\sigma_r/b$ does not intersect the square, the investor cannot take all that additional interest rate risk which he wants to take. Hence, the top right-hand corner of the square represents the optimal combination of elasticity and duration.

As before, we proceed by computing the optimal portfolio processes.

Corollary 5.4 (Optimal portfolios in the Briys-de Varenne model)
(i) Assume that elasticity and duration meet the condition

$$\varepsilon_S(t)D_B(t) - \varepsilon_B D_S(t) \neq 0 \tag{5.46}$$

at time $t \in [0, \tau]$. If the investor can trade in stocks and defaultable bonds, then the optimal portfolio process $(\pi_S^*(t), \pi_B^*(t))$ at time t is given by

$$\pi_S^*(t) = \frac{\varepsilon^*(t)D_B(t) - \varepsilon_B(t)D^*(t)}{\varepsilon_S(t)D_B(t) - \varepsilon_B(t)D_S(t)}, \qquad \pi_B^*(t) = \frac{\varepsilon_S(t)D^*(t) - \varepsilon^*(t)D_S(t)}{\varepsilon_S(t)D_B(t) - \varepsilon_B(t)D_S(t)}.$$

Otherwise, there exists a constant $c(t) \in \mathbb{R}$ *such that portfolio duration can be expressed as a muliple of portfolio elasticity, i.e.*

$$\pi_S(t)D_S(t) + \pi_B(t)D_B(t) = c(t)\Big(\pi_S(t)\varepsilon_S(t) + \pi_B(t)\varepsilon_B(t)\Big). \tag{5.47}$$

Besides, only the elasticity $\tilde{\varepsilon}(t)$ *of Proposition 5.8 with* $R(t) = c(t)$ *is attainable and each portfolio process* $(\pi_S^*(t), \pi_B^*(t))$*, which leads to this elasticity, is optimal, i.e. the optimal portfolio process is not uniquely determined.*

(ii) If the investor can only trade in defaultable bonds, then the optimal portfolio process is given by

$$\pi_B^*(t) = \frac{\tilde{\varepsilon}(t)}{\varepsilon_B(t)}.$$

Proof.

(i) To compute the optimal portfolio process we solve the following system of equations

$$\begin{aligned} \varepsilon^*(t) &= \pi_S(t)\varepsilon_S(t) \; + \pi_B(t)\varepsilon_B(t), \\ D^*(t) &= \pi_S(t)D_S(t) + \pi_B(t)D_B(t). \end{aligned}$$

If condition (5.46) is met, the determinant of the matrix

$$A(t) := \begin{pmatrix} \varepsilon_S(t) & \varepsilon_B(t) \\ D_S(t) & D_B(t) \end{pmatrix}$$

is not zero. Hence, the inverse of A exists, which leads to

$$(\pi_S^*(t), \pi_B^*(t))' = A(t)^{-1}(\varepsilon^*(t), D^*(t))'.$$

If condition (5.46) is not satisfied, the row vectors of the matrix are linearly dependent, which yields (5.47). As a consequence, the investor cannot choose elasticity and duration independently. Optimizing over elasticity gives the same result as in Proposition 5.8 with $R(t) = c(t)$. Note that $c(t)$ is equal to the ratio of duration and elasticity. Clearly, the optimal portfolio process is not uniquely determined because the investor can put funds into two corporate securities to match the elasticity $\tilde{\varepsilon}$.

(ii) To match the elasticity $\tilde{\varepsilon}$ the investor chooses the fraction π_B so that $\tilde{\varepsilon}(t) = \pi_B(t) \cdot \varepsilon_B(t)$. This gives the result. □

5.5 Conclusion

In this chapter we have considered portfolio problems with defaultable securities. Default risk has been modeled using the firm value framework of Merton (1974), Black/Cox (1976), and Briys/de Varenne (1997). As in this framework the price of a corporate security is given by the price of a contingent claim on firm value, we have actually faced a portfolio problem with derivatives. Apart from this, an additional restriction comes into play which can be interpreted as accounting equation. This is due to the fact that in firm value approaches the capital structure of a firm is explicitly modeled. Consequently, the investor can buy at most one share of equity and one share of debt, which leads to stochastic bounds on the investor's portfolio process. Actually, these bounds depend on the control itself. Using a "small investor assumption" we have first ignored the accounting equation and solved the resulting portfolio problem in the Merton model and the Black-Cox model. However, our approach can be applied to any firm value model where corporate securities are contingent claims on firm value modeled by a geometric Brownian motion. Besides, it is valid for a broad class of utility functions. This is due to the fact that we have used the elasticity approach to portfolio optimization which has proved to be very helpful when derivatives belong to the investment opportunities. Whereas in the Merton model the optimal portfolio processes are fairly easy to compute because the market is complete, in the Black-Cox model a stopped portfolio problem needs to be solved.

In a next step we have investigated the corresponding constrained portfolio problems. We have restricted our considerations to an investor who maximizes utility from terminal wealth with respect to a logarithmic utility function. The reason for this assumption is that bounds on the investor's portfolio process are difficult to handle. Although we have focused on the models of Merton (1974), Black/Cox (1976), and a generalized version of Briys/de Varenne (1997) - in contrast to Briys/de Varenne (1997) we have used a general one-factor model for the short rate - our approach can be applied to other firm value models such as the models by Geske (1977), Longstaff/Schwartz (1995), or Saa-Requejo/Santa-Clara (1999). The study of the constrained problem for more general utility functions will form a topic for future research.

There exists another approach for modeling default risk known as reduced-form approach, which has been developed in Jarrow/Turnbull (1995).[18] Formally, default is triggered when a jump of a Poisson process occurs, which is usually not linked to firm value. Therefore, there exists no explicit connection between default and capital structure of the firm. Hence, if we model default by a reduced-form model, a portfolio problem with defaultable securities can be solved similarly to an ordinary stopped portfolio problem in a jump-diffusion framework. To consider this framework in more detail will also form another topic for future research.

[18] See e.g. Jarrow/Turnbull (2000) for a survey of credit risk models.

References

Amin, K. I.; R. A. Jarrow (1991): Pricing foreign currency options under stochastic interest rates, *Journal of International Money and Finance* 10, 310-329.

Amin, K. I.; R. A. Jarrow (1992): Pricing options on risky assets in a stochastic interest rate economy, *Mathematical Finance* 2, 217-237.

Baxter, M. (1997): Generel interest rate-models and the universality of HJM, in: Dempster, M. A. H.; S. R. Pliska, eds., *Mathematics of Derivative Securities*, Cambridge University Press, Campridge, 315-335.

Billingsley, P. (1986): *Probability and Measure*, 2nd ed., John Wiley & Sons, New York.

Björk, T. (1998): *Arbitrage theory in continuous time*, Oxford University Press, Oxford.

Black, F. (1976): The Pricing of commodity contracts, *Journal of Financial Economics* 3, 167-179.

Black, F.; J. C. Cox (1976): Valuing corporate securities: Some effects of bond indenture provisions, *Journal of Finance* 31, 351-367.

Black, F.; E. Derman; W. Toy (1990): A one-factor model of interest rates and its application to treasury bond options, *Financial Analysts Journal* 46, 33-39.

Black, F.; P. Karasinski (1991): Bond and option pricing when the short rates are lognormal, *Financial Analysts Journal* 47, 52-59.

Black, F.; M. Scholes (1973): The Pricing of options and corporate liabilities, *Journal of Political Economy* 81, 637-654.

Bowie, J.; P. Carr (1994): Static simplicity, *Risk* 7, August, 45-49.

Briys, E.; F. de Varenne (1997): Valuing risky fixed rate debt: An extension, *Journal of Financial and Quantitative Analysis* 32, 239-248.

Canestrelli, E. (1998): Inquiries on the applications of multidimensional stochastic processes to financial investments, *Working Paper*, Dipartimento di Matematica Applicata ed Informatica, Universita Ca' Foscari, Venezia.

Carr, P. (1995): Two extensions to barrier option valuation, *Applied Mathematical Finance* 2, 173-209.

Collin-Dufresne, P.; R. Goldstein (2001): Do credit spreads reflect stationary ratios?, *Journal of Finance* 56, 1929-1957.

Collin-Dufresne, P.; R. Goldstein; S. Martin (2001): The determinants of credit spread changes, *Journal of Finance* 56, 2177-2208.

Cox, J. C.; C. F. Huang (1989): Optimal consumption and portfolio policies when asset prices follow a diffusion process, *Journal of Economic Theory* 49, 33-83.

Cox, J. C.; C. F. Huang (1991): A variational problem arising in financial economics, *Journal of Mathematical Economics* 20, 465-487.

Cox, J. C.; J. E. Ingersoll, Jr.; S. A. Ross (1979): Duration and the measurement of basis risk, *Journal of Business* 52, 51-62.

Cox, J. C.; J. E. Ingersoll, Jr.; S. A. Ross (1981): The relation between forward prices and futures prices, *Journal of Financial Economics* 9, 321-346.

Cox, J. C.; J. E. Ingersoll, Jr.; S. A. Ross (1985): A theory of the term structure of interest rates, *Econometrica* 53, 385-407.

Cox, J.; M. Rubinstein (1985): *Option markets*, Prentice Hall, Englewood Cliffs, N.J.

Cvitanic, J.; I. Karatzas (1992): Convex duality in constrained portfolio optimization, *Annals of Applied Probability* 2, 767-818.

Deelstra, G.; M. Grasselli; P.-F. Koehl (2000): Optimal investment strategies in a CIR framework, *Journal of Applied Probability* 37, 936-946.

Deelstra, G.; M. Grasselli; P.-F. Koehl (2002): Optimal investment strategies in the presence of a minimum guarantee, *Working Paper*, RUG, Gent.

Dothan, M. (1978): On the term structure of interest rates, *Journal of Financial Economics* 7, 229-264.

Dybvig, P. H.; L. C. G. Rogers; K. Back (1999): Portfolio turnpikes, *Review of Financial Studies* 12, 165-195.

Duffie, D. (1992): *Dynamic asset pricing theory*, Princeton University Press, Princeton.

Duffie, D.; R. Kan (1996): A yield factor model of interest rates, *Mathematical Finance* 6, 379-406.

Elliott, R. J.; P. E. Kopp (1999): *Mathematics of financial markets*, Springer, New York.

Eom, Y. H.; J. Helwege; J.-z. Huang (2002): Strutural models of corporate bond pricing: An empirical analysis, *Working Paper*, Ohio State University, Columbus.

Ericsson, J; J. Reneby (1998): A framework for valuing corporate securities, *Applied Mathematical Finance* 5, 143-163.

Feller, W. (1951): Two singular diffusion problems, *Annals of Mathematics* 54, 173-182.

Fleming, W. H.; R. W. Rishel (1975): *Deterministic and stochastic optimal control*, Springer, New York.

Fleming, W. H.; H. M. Soner (1993): *Controlled Markov processes and viscosity solutions*, Springer, New York.

Geman, H.; N. El Karoui; J.C. Rouchet (1995): Changes of numeraire, changes of probability measure and option pricing, *Journal of Applied Probability* 32, 443-458.

Geske, R. (1977): The valuation of corporate liabilites as compound options, *Journal of Financial and Quantitative Analysis* 12, 541-552.

Gihman, I. L.; A. V. Skorohod (1972): *Stochastic differential equations*, Springer, New York.

Goldman, M.; H. Sosin; M. Gatto (1979): Path dependent options: "Buy at the low and sell at the high", *Journal of Finance* 34, 1111-1126.

Harrison, J.M. (1985): *Brownian motion and stochastic flow systems*, Wiley, New York.

Harrison, J. M.; D. M. Kreps (1979): Martingales and arbitrage in multiperiod securities markets, *Journal of Economic Theory* 20, 381-408.

Harrison, J. M.; S. R. Pliska (1981): Martingales and stochastic integrals in the theory of continuous trading, *Stochastic Processes and Applications* 11, 215-260.

Harrison, J. M.; S. R. Pliska (1983): A stochastic calculus model of continuous trading: complete markets, *Stochastic Processes and Applications* 15, 313-316.

Heath, D., R. Jarrow, A. Morton (1992): Bond pricing and the term structure of interest rates: A new methodology for contingent claim valuation, *Econometrica* 60, 77-105.

Ho, T. S. Y., S.-B. Lee (1986): Term structure movements and pricing interest contingent claims, *Journal of Finance* 41, 1011-1029.

Hogan, M.; K. Weintraub (1993): The lognormal interest rate model and Eurodollar futures, *Working Paper*, Citibank, New York.

Hull, J.; A. White (1990): Pricing interest-rate derivative securities, *Review of Financial Studies* 3, 573-592.

Hull, J. C.; A. White (1995): The impact of default risk on the prices of options and other derivative securities, *Journal of Banking and Finance* 19, 299-322.

Ikeda, N., S. Watanabe (1981): *Stochastic differential equations and diffusion processes*, North Holland, New York.

Jarrow, R. A.; S. M. Turnbull (1995): Pricing derivatives on financial securities subject to credit risk, *Journal of Finance* 50, 53-85.

Jarrow, R. A.; S. M. Turnbull (2000): The intersection of market and credit risk, *Journal of Banking and Finance* 24, 271-299.

Johnson, H.; R. Stulz (1987): The pricing of options with default risk, *Journal of Finance* 42, 267-280.

Jones, E.; S. Mason; E. Rosenfeld (1984): Contingent claims analysis of corporate capital structures: An empirical investigation, *Journal of Finance* 39, 611-627.

Karatzas, I.; J. P. Lehoczky; S. E. Shreve (1987): Optimal portfolio and consumption decisions for a small investor on a finite horizon, *SIAM Journal on Control and Optimization* 27, 1157-1186.

Karatzas, I.; S. E. Shreve (1991): *Brownian motion and stochastic calculus*, 2nd ed., Springer, New York.

Karlin, S.; H. M. Taylor (1981): *A second course in stochastic processes*, Academic Press, New York.

Kim, J.; K. Ramaswamy; S. Sundaresan (1993): Does default risk in coupons affect the valuation of corporate bond?: A contingent claims model, *Financial Management*, 117-131.

Kolkiewcz, A.W. (2002): Pricing and hedging more general double-barrier options, *Journal of Computational Finance* 5, 1-26.

Korn, R. (1997): *Optimal portfolios*, World Scientific, Singapore.

Korn, R.; E. Korn (2001): *Option pricing and portfolio optimization - Modern methods of financial mathematics*, AMS, Providence, Rhode Island.

Korn, R.; H. Kraft (2001): A stochastic control approach to portfolio problems with stochastic interest rates, *SIAM Journal on Control and Optimization* 40, 1250-1269.

Korn, R.; H. Kraft (2003): Optimal portfolios with defaultable securities - A firm value approach, *International Journal of Theoretical and Applied Finance* 6, 793-819.

Korn, R.; S. Trautmann (1999): Optimal control of option portfolios, *OR-Spektrum* 21, 123-146.

Kraft, H. (2003): Curved barriers and default, *Wilmott magazine*, July issue, 68-73.

Kraft, H. (2003): Elasticity approach to portfolio optimization, *Mathematical Methods of Operations Research* 58, 159-182.

Krylov, N. (1980): *Controlled diffusion processes*, Springer, Berlin.

Kunitomo, N.; M. Ikeda (1992): Pricing options with curved boundaries, *Mathematical Finance* 2, 275-298.

Leland, H. (1994): Corporate debt value, bond covenants, and optimal capital structure, *Journal of Finance* 49, 1213-1252.

Leland, H.; K. Toft (1996): Optimal capital structure, endogenous bankruptcy, and the term structure of credit spreads, *Journal of Finance* 51, 987-1019.

Lioui, A.; P. Poncet (2001): On optimal portfolio choice under stochastic interest rates, *Journal of Economic Dynamics and Control* 25, 1841-1865.

Longstaff/Schwartz (1995): A simple approach to valuing risky fixed and floating rate debt, *Journal of Finance* 50, 789-819.

Magshoodi, Y. (1996): Solution of the extended CIR term structure and bond option valuation, *Mathematical Finance* 6, 89-109.

Markowitz, H. M. (1952): Portfolio selection, *Journal of Finance* 7, 77-91.

Mason, S. P. ; S. Bhattacharya (1981): Risky debt, jump processes, and safety covenants, *Journal of Financial Economics* 9, 281-307.

Merton, R. C. (1969): Lifetime portfolio selection under uncertainty: the continuous case, *Reviews of Economical Statistics* 51, 247-257.

Merton, R. C. (1971): Optimal consumption and portfolio rules in a continuous-time model, *Journal of Economic Theory* 3, 373-413. Erratum: ebenda 6 (1973), 213-214.

Merton, R. C. (1973): The theory of rational option pricing, *Bell Journal of Economics and Management Science* 4, 141-183.

Merton, R. C. (1974): On the pricing of corporate debt: The risk structure of interest rates, *Journal of Finance* 29, 449-479.

Merton, R. C. (1990): *Continuous-time finance*, Basil Blackwell, Cambridge MA.

Munk, C. (1999): Stochastic duration and fast coupon bond option pricing in multifactor models, *Review of Derivatives Research* 3, 157-181.

Musiela, M.; M. Rutkowski (1997): *Martingale methods in financial modelling*, Springer, Berlin.

Nielsen, L. T. (1999): *Pricing and hedging of derivative securities*, Oxford University Press, Oxford.

Novikov, A,; V. Frishling; N. Kordzakhia (1999): Approximations of boundary crossing probabilities for a Brownian motion, *Journal of Applied Probability* 36, 1019-1030.

Pitman, J.W.; M. Yor (1982): A decomposition of Bessel bridges, *Zeitschrift für Wahrscheinlichkeitstheorie und verwandte Gebiete* 59, 425-457.

Pliska, S. R. (1986): A stochastic calculus model of continuous trading: Optimal portfolios, *Mathematics of Operations Research* 11, 371-382.

Roberts, G. O.; C. F. Shortland: Pricing barrier options with time-dependent coefficients, *Mathematical Finance* 7, 83-93.

Rogers, L. C. G.; O. Zane. (1997): Value moving barrier options, *Journal of Computational Finance* 1, 5-11.

Rubinstein, M.; E. Reiner (1991): Breaking down the barriers, *RISK* 4, 28-35.

Saa-Requejo, J.; Santa-Clara, P. (1999): Bond pricing with default risk, *Working paper*, UCLA.

Sandmann, K.; D. Sondermann (1997): A note on the stability of lognormal interest rate models and the pricing of Eurodollar futures, *Mathematical Finance* 7, 119-125.

Sethi, S.; M. Taksar (1988): A note on Merton's "Optimal consumption and portfolio rules in a continuous-time model", *Journal of Economic Theory* 46, 395-401.

Shirakawa (2002), H.: Squared Bessel processes and their applications to the square root interest rate model, *Asia-Pacific Financial Markets* 9, 169-190.

Shreve, S. (1997): *Stochastic calculus and finance*, Lecture notes, Carnegie Mellon University.

Skorokhod, A. V. (1965): *Studies in the theory of random processes*, Addison-Wesley, Reading, Massachusetts.

Sørensen, C. (1999): Dynamic asset allocation and fixed income management, *Journal of Financial and Quantitative Analysis* 34, 513-531.

Vasicek, O. (1977): An equilibrium characterization of the term structure, *Journal of Financial Economics* 5, 177-188.

Wachter, J. A. (2001): Risk aversion and allocation to long-term bonds, *Working Paper*, New York University, Stern School of Business, New York.

Yamada, T. (1981) On the successive approximation of solutions of stochastic differential equations, *Journal of Mathematics of Kyoto University* 21, 501-515.

Yamada, T.; S. Watanabe (1971): On the uniquenes of solutions of stochastic differential equations, *Journal of Mathematics of Kyoto University* 11, 155-167.

Abbreviations

a.s.:	almost surely
cond.:	condition
const.:	constant
CSDE:	controlled stochastic differential equation
determ.:	deterministic
ed.:	1. edition 2. editor
e.g.:	for example
EAPO:	elasticity approach to portfolio optimization
f:	following page
ff:	following pages
i.e.:	that is
HJB:	Hamilton-Jacobi-Bellman equation
KTC:	Kuhn-Tucker condition
LP:	liquidity premium
p.:	page
ODE:	ordinary differential equation
PDE:	partial differential equation
RP:	risk premium
SDE:	stochastic differential equation
syn.:	synonym
stoch.:	stochastic
w.r.t.:	with respect to

Notations

$\| \|$:	Euklidian norm or operator norm
$\| \|_\infty$:	row-sum norm
$a \wedge b$:	minimum of the real numbers a and b
$\bar{U}$:	closure of the set U
M':	transpose of the matrix M
arg:	argument
$\mathcal{B}(U)$:	Borel-σ-algebra over the set U
$C^{1,2}$:	space of the continuous functions which are continuously differentiable with respect to the first variable and two-times continuously differentiable with respect to the second variable
∂U:	boundary of the set U
$\text{diag}(a_1, \ldots, a_n)$:	diagonal matrix with diagonal elements $a_1, \ldots, a_n$
$\text{diam}(U)$:	diameter of the set U
exp:	exponential function
$\text{e}(X)$:	expected value of the random variable X
$\inf(U)$:	infimum of the set U
ln:	logarithm to base e
$I\!N$:	natural numbers
$\max(U)$:	maximum of the set U
$\min(U)$:	minimum of the set U
$I\!R$:	real numbers
$I\!R_+$:	$\{y \in I\!R : y > 0\}$
$I\!R_+^0$:	$I\!R_+ \cup \{0\}$
$\sup(U)$:	supremum of the set U

Given some function $f \in C^{1,2}$ depending on t and x and some stochastic process Z, throughout this thesis f_t, f_x, f_{xx} denote the partial derivatives of f and $Z(t)$ denotes the value of Z at time t.

Druck und Bindung: Strauss GmbH, Mörlenbach